# An Introduction to Sustainable Development

WITHDRAWN

'Jennifer Elliott's treatment of sustainability is concise and well illustrated . . . the reader is acquainted with a broad range of contemporary themes and provocative concepts. It is a valuable addition to the introductory literature.'

*Progress in Human Geography*, review of first edition

Recent developments and interventions in the field of sustainable development have made this revised and updated second edition of *An Introduction to Sustainable Development* particularly timely as the world rises to the challenge of determining the prospects of sustainable development in the future.

The book evaluates the progress made in the last decades of the twentieth century towards establishing new patterns and processes of development which are sustainable; in terms of the demands they make on the world's physical, ecological and cultural resources and the characteristics of technology, societal organisation and economic production which underpin them.

The focus of the book is the developing world, where conditions such as rising poverty and mounting debt combine to present particular challenges of sustainability. The resolution of these issues is seen to depend on changes that include the operation of international institutions, the activities of government and private sector interests, as well as changes in the behaviour of community organisations and individuals across the globe.

Containing a wealth of student-friendly features including boxed case studies drawn from across the world, discussion questions, guides for further reading and a glossary, this text provides an invaluable introduction to the characteristics, challenges and opportunities of sustainable development.

**Jennifer A. Elliott** is a Senior Lecturer in Geography at the University of Brighton.

# In the same series

**Routledge Introductions to Development Series**

# An Introduction to Sustainable Development

## Second edition

Jennifer A. Elliott

London and New York

First published 1994
by Routledge
11 New Fetter Lane, London EC4P 4EE

Simultaneously published in the USA and Canada
by Routledge
29 West 35th Street, New York, NY 10001

*Routledge is an imprint of the Taylor & Francis Group*

Reprinted 1996
Second edition first published 1999
© 1994, 1999 Jennifer A. Elliott

The right of Jennifer A. Elliott to be identified as the Author
of this Work has been asserted by her in accordance with the
Copyright, Designs and Patents Act 1988

Typeset in Times by RefineCatch Ltd, Bungay, Suffolk
Printed and bound in Great Britain by
Biddles Ltd, Guildford and King's Lynn

*British Library Cataloguing in Publication Data*
A catalogue record for this book is available from the British Library

*Library of Congress Cataloging in Publication Data*
Elliott, Jennifer A., 1962–
 An Introduction to Sustainable Development/Jennifer A. Elliott.
 – 2nd ed.
  p. cm. – (Routledge Introductions to Development Series)
  Includes bibliographical references and index.
  1. Sustainable development—Developing countries.
  2. Environmental policy—Developing countries.  I. Title.
  II. Series: Routledge Introductions to Development.
  HC59.72.E5E 43  1999
  338.9′009172′4—dc21    98–54351 CIP

ISBN 0–415–19150–5 (hbk)
ISBN 0–415–19151–3 (pbk)

*For Lesley and Thomas*

# Contents

# Plates

# Figures

# Tables

# Boxes

# Acknowledgements

I am very grateful to many people who supported me in writing this book. In particular, to Becky Elmhirst for her expertise and humour, and to Kirsty Smallbone for many of the diagrams. Thanks also to colleagues (past and present) who kindly lent me slides and to Casey Mein at Routledge for her help in various forms. My biggest thanks are to my family for their sustained interest and tolerance.

# Introduction

This book is concerned with the challenges and opportunities of finding sustainable patterns and processes of development within the international community for the future. It is now widely recognised by academics and practitioners in many fields, both in the developing and more industrialised countries, that development to date has too regularly led to the degradation of resources, for example. Mounting world poverty is also testimony to the failure of such economic and social transformations to deliver even basic goods to many people (particularly in the developing regions). It is now suggested that these patterns and processes of development will not be able to supply the needs of the world's population into the future and cannot deliver the higher standards of living to the rising numbers of people essential to the conservation of the environment.

One of the primary aims of the book is to highlight the progress made in the last decades of the twentieth century towards establishing new patterns and processes of development which are more sustainable: in terms of the demands they make on the physical, ecological and cultural resources of the globe, and the characteristics of technology, societal organisation and economic production which underpin them. Understanding the characteristics of successful sustainable development projects will be essential for meeting the ongoing and evolving challenges world-wide of balancing present needs against those of the future.

It will be seen that the prospects of sustainable development in any one location are in part shaped by forces and decision-making which are often situated at great distances away. It is therefore impossible to consider the developing world in isolation from the wider global community. However, there are also particular and distinct issues of sustainability in the developing world, and these challenges and opportunities are the major focus of this book.

Questions regarding the fostering of sustainable development will also be seen to lie in both the natural and the human environment. For example, many countries of what can be termed the developing world are in the tropics. The tropical ecosystem is a fragile environment that is easily disturbed (see Gupta, 1998) and sets boundary conditions on development, particularly in agriculture, which are often quite different to those of temperate regions. Large sections of the populations of these countries live in physical environments in which securing basic needs is extremely problematic and which may even be detrimental to human health. Indeed, premature deaths and rising numbers of people in poverty are characteristics of the human environment in many areas of the developing world (to an extent not seen outside the region), which further combine to create particular challenges and opportunities of sustainable development.

However, in order to understand the characteristics of resource use or human condition in the developing world and to allow more sustainable patterns to be supported, it is essential to identify the underlying processes of change. Some of these processes may operate solely at a local level, whilst others may impact across many places and constitute global forces of change. All to some degree, and in combination, shape the interactions between people and the environment (wherever they live) and the relationships between people in different places. It is for these reasons that sustainable development is a common challenge for the global community as a whole. In the course of this book, it will be seen that sustainable development in the future requires actions for change at all levels, addressing both the human and physical environments, through interventions in physical, political-economic and social processes.

In Chapter 1, the origins of this apparent common global agenda of sustainable development in the late 1990s are traced within an analysis of key ideas in development theory and landmarks in environmentalism. Whilst the interdependence of future environment and development ends are recognised in both literatures, the context within which they are being pursued is changing rapidly, requiring constant re-evaluation, including of the relationship between nations at differing levels of economic development in different parts of the globe.

In Chapter 2, the impacts of past development processes on both people and the environments of the world are discussed in detail, providing a fuller insight into the nature of the challenges of sustainable development for individual actors and the various institutions of development. Issues of the interdependence of the developing and developed worlds are again to the fore in the consideration of questions of responsibility for past environmental degradation and managing change into the future, for example.

In Chapter 3, the range of actions which have been taken at a variety of levels towards ensuring sustainable development in the future are identified. The ways in which people and places across the globe are interconnected are made explicit through consideration of the major issues of trade, aid and debt. The increasing and varied actions of non-governmental organisations (NGOs), particularly those focused on empowering local communities, are also highlighted. Whilst it is evident that many institutions of development are transforming what they do, the critical challenge of translating the substantial rhetoric which surrounds sustainable development into practical actions is identified as resting on the ways in which these actors at various scales *work together* in the future.

In Chapters 4 and 5, the particular challenges and opportunities of sustainable development in the developing world are considered in rural and urban contexts. It is quickly seen that the two sectors are not distinct and that the environment and development concerns therein are often interrelated. Furthermore, the principles which are seen to be now guiding more sustainable development interventions in practice are regularly common to both rural and urban settings. For example, addressing the welfare needs of the poorest groups and building responsive and inclusive systems of research and development are identified as being essential to achieving the goals of development and conservation in either sector. However, important differences are also seen in rural and urban areas in terms of the nature of the immediate environmental problems and development concerns, the options for securing income and livelihood, the hazards and sources of instability of living and working in these sectors, and the specific opportunities for action.

In Chapter 6, the ongoing challenges and opportunities of sustainable development for the global community are summarised. An assessment is made of the prospects of further change at the turn of the century in the light of the continued tensions but also the significant changes in practice which have been identified in the substantive chapters of the book.

# What is sustainable development?

- Definitions of the concept of sustainable development
- Changing ideas about development
- Key landmarks in environmentalism
- The context for sustainable development in the late 1990s
- Globalisation

## Introduction

In 1992, the United Nations Conference on Environment and Development, the 'Earth Summit', took place in Rio de Janeiro, Brazil. It was the largest ever international conference and the central aim was to identify the principles of action towards 'sustainable development' in the future. The challenge was seen to require consensus at the highest level, such that, for the first time, heads of state gathered to consider the environment.

By the late 1990s, the term sustainable development has 'gained a currency well beyond the confines of global environmental organisations' (Adams, 1990: 2) and is widely used in many political arenas and academic fields. Certainly in the developed world, the substantial media attention given to the serious environmental disturbances surrounding forest fires in Indonesia, flooding in the Americas, China and Bangladesh, and typhoons in South-East Asia, for example, has brought questions of conservation and ideas of sustainability into the public vocabulary. In the fields of development and the environment, there is now an evident consensus that sustainable development is an important rallying point for research and action and a desirable policy objective which should be striven for.

Whilst the primary output of the UN Conference on Environment and Development, the huge Agenda 21 document, carries much political authority and moral force (Mather and Chapman, 1995) towards reconciling conservation and development actions into the twenty-first century, substantial debate over the meaning and practice of sustainable development continues. Important tensions persist, for example, between the environmental concerns of rich and poor countries, between those who wish to exploit resources and those who wish to conserve them, and between the development needs of current generations and those of the future.

This chapter identifies in some detail the origins of the concept of sustainable development and its 'meaning' at the end of the twentieth century in terms of finding alternative patterns of progress to meet the needs of the global community. Through an analysis of the key debates in the previously separate literatures of development thinking and environmentalism, it is possible to understand the sources of continued conflict regarding sustainable development in theory and practice and the broad political economic context in which sustainable development is being sought.

## The concept of sustainable development

Literally, sustainable development refers to maintaining development over time. However, it has been suggested that there are more than seventy definitions of sustainable development currently in circulation (Holmberg and Sandbrook, 1992). Figure 1.1 lists just a small number of such definitions and the varied interpretations of the concept which have flowed from these different ideas. Definitions are important, as they are the basis on which the means for achieving sustainable development in the future are built. During the course of this text, it will be apparent that, although there are many signs of progress, there is also much uncertainty as to the most appropriate strategies to foster sustainable change. Indeed, as suggested in the quotations in Figure 1.1, the attractiveness (and the 'dangers') of the concept of sustainable development may lie precisely in the varied ways in which it can be interpreted and used to support a whole range of interests or causes.

The challenges of understanding what this idea of sustainable development may mean, and how people can work towards it, are evident in a brief analysis of the most widely cited definition of sustainable development, that of the World Commission on Environment and Development (WCED). This Commission was formed by the United Nations in 1984. It was an independent group of 22 people drawn from member states of both the developing and

**Figure 1.1**  *Defining and interpreting the contested concept of sustainable development*
....................................................................................................

**Definitions of sustainable development**

'In principle, such an optimal (sustainable growth) policy would seek to maintain an "acceptable" rate of growth in per-capita real incomes without depleting the national capital asset stock or the natural environmental asset stock.'                                (Turner, 1988: 12)

'The net productivity of biomass (positive mass balance per unit.area per unit time) maintained over decades to centuries.'                                (Conway, 1987: 96)

'Development that meets the needs of the present without compromising the ability of future generations to meet their own needs.'
                                (World Commission on Environment and Development, 1987: 43)

**Interpretations of sustainable development**

'A creatively ambiguous phrase . . . an intuitively attractive but slippery concept.'
                                (Mitchell, 1997: 28)

'Like motherhood, and God, it is difficult not to approve of it. At the same time, the idea of sustainable development is fraught with contradictions.'
                                (Redclift, 1997: 438)

'It is indistinguishable from the total development of society.'                (Barbier, 1987: 103)

'Its very ambiguity enables it to transcend the tensions inherent in its meaning.'
                                (O'Riordan, 1995: 21)
....................................................................................................

developed worlds, charged with identifying long-term environmental strategies for the international community. The apparently simple definition of sustainable development forwarded by the Commission (as reproduced in Figure 1.1) is soon seen to encompass some very challenging notions, such as those of inter-generational equity, needs and limits. For example, such questions emerge as: what is it that one generation is passing to another? Is it solely natural capital or does it include assets associated with human ingenuity or culture? What and how are the limits set – by technology, society or ecology, for example? What of the fact that, currently, needs in one place or amongst particular groups are often fulfilled at the expense of others, such as those in the developing world? Fundamentally, 'needs' mean different things to different people and are linked to our ability to satisfy them, i.e. are closely aligned to 'development' itself. So, society is able to define and create new 'needs' within certain groups, without satisfying even the basic needs of others.

**Plate 1.1** *Promoting the messages of sustainable development*
   *a. Sign on entry to Kang, Botswana*
   *b. VOYCE (Views of Young Concerned Environmentalists)* **Four Seasons mural, Brighton, England**

Source: David Nash, University of Brighton.

(a)

(b)

Source: Kim Jackson, Brighton and Hove council.

**Figure 1.2** *Core issues and necessary conditions for sustainable development as identified by the World Commission on Environment and Development (WCED)*

........................................................................................................................

**Core issues:**
- Population and development
- Food security
- Species and ecosystems
- Energy
- Industry
- The urban challenge

**Pursuit of sustainable development requires:**

A political system that secures effective citizen participation in decision-making

An economic system that provides for solutions for the tensions arising from disharmonious
    development

A production system that respects the obligation to preserve the ecological base for
    development

A technological system that fosters sustainable patterns of trade and finance

An international system that fosters sustainable patterns of trade and finance

An administrative system that is flexible and has the capacity for self-correction

........................................................................................................................

Source: WCED (1987).

Furthermore, the substantial challenges of operationalising the concept of sustainable development were clear in the report of the WCED. Figure 1.2 displays the core issues identified by the Commission and the necessary conditions for sustainable development in the future. A more prosperous, more just and more secure global future was seen to depend on new norms of behaviour at all levels and in the interests of all. The conditions for such a future encompass all areas of human activity, in production, trade, technology and politics, for example, and encompass co-operative and mutually supportive actions on behalf of individuals and nations at all levels of economic development.

Clearly, whilst common sense would seem to tell us that our development should not be at the expense of that of future generations, the challenges in practice are substantial. In order to identify the challenges of implementing sustainable development actions and to realise the opportunities for sustainable development, it is necessary to understand the changes in thinking and practice from which the concept has developed. As Adams (1990) suggests, sustainable development cannot be understood in 'an historical vacuum' (p. 14). Of particular importance are the changes in thinking about what constitutes 'development' and how best to achieve it, and changing ideas about the 'environment'.

## Changing perceptions of development

Development is often discussed in relation to 'developing countries', but in fact, it is a concept which relates to all parts of the world at every level, from the individual to global transformations (Potter *et al.*, 1999). 'Development' is therefore something to which we all aspire. Ideas about the best means by which to achieve our aspirations and needs are potentially as old as human civilisation. The study of development, however, has a relatively short history, really dating back only as far as the 1950s. Since then, the interdisciplinary field of development studies has seen many changes in thinking regarding the meaning and purpose of development (ideologies) and in development practice in the field (strategies of development). Although these shifts are considered chronologically here, in reality existing theories are rarely totally replaced; rather new ones find relative favour and emphasis.

During the first United Nations Development Decade of the 1960s, development thinking (encompassing these aspects of ideology and strategy) prioritised economic growth and the application of modern scientific and technical knowledge as the route to prosperity in the underdeveloped world at that time. It was a period characterised by optimism and global co-operation, within which it was assumed that many development problems of the underdeveloped world would be solved quickly through the transfer of finance, technology and experience from the developed countries. Insights from neo-classical economics as modelled by authors such as Hirschmann (1958) and Rostow (1960) were very influential in development thinking at this time. Rostow's model of the linear stages of economic development is shown in Figure 1.3. On the basis largely of the experience of the more developed societies, it was suggested that, through assistance in reaching a critical 'take-off' stage in levels of savings and investment, the benefits of development and characteristics of 'modernisation' would inevitably and spontaneously flow from the core to less-developed regions. Because of the way development was modelled as becoming 'more like the West' through processes of spatial diffusion, such thinking has been referred to as 'modernisation theory'.

However, by the second UN Development Decade of the 1970s,

**Figure 1.3**  *The stages of economic development as modelled by Rostow (1960)*

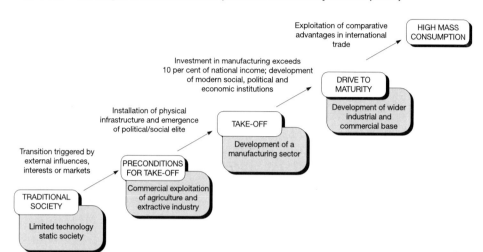

inequality between and within countries had in fact worsened. Many developing countries had achieved economic growth as measured by Gross National Product (GNP) but this 'development' was not shared equally amongst the populations of these nations. For example, in Brazil in 1970, the poorest 40 per cent of the population received only 6.5 per cent of the total national income, in contrast to the 66.7 per cent of the total national income received by the richest 20 per cent of the population (Todaro, 1997). The optimism of a speedy end to underdevelopment faded.

During the 1970s, development thinking was influenced strongly by the writings of scholars within the developing world itself. Significantly, their ideas related to the conditions of those countries rather than arising through ideas emanating from conditions in Europe (Potter *et al.*, 1999). Their work became known as the radical or 'dependency' school of thought and included the work of Andre Gunder Frank (1967). Fundamentally, the assertion was that underdevelopment was the direct outcome of development elsewhere and the operations of the international capitalist system. To use Frank's terminology, development and underdevelopment are two sides of the same coin. As illustrated in Figure 1.4, peripheral or satellite regions and countries are integrated into the world system through processes of unequal exchange and dependent relations with the metropolitan core. In consequence, the further entrenched they become in such processes, the more they are held back in development, rather than enabled to progress. The barriers to

**Figure 1.4** *The Frank model of underdevelopment*

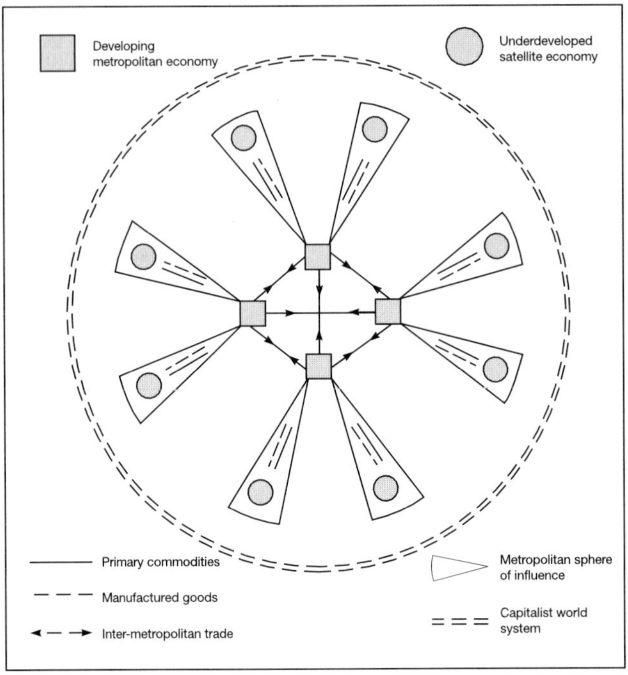

Source: Corbridge (1987: 20–2).

development, therefore, as modelled by dependency theorists, lay in the international division of labour rather than a lack of capital or entrepreneurial skills, as within modernisation thinking.

By the third UN Development Decade of the 1980s, distributional issues, such as improving the income levels of target populations, were accepted as fundamental parts of any development strategy. Phrases such as 'growth with equity' or 'redistribution with growth' emerged in the 1970s and encapsulated the recognition that economic growth remains a fundamental ingredient within development thinking and action, but that the nature of that growth was critical to ensuring that the benefits do not fall solely to a minority of the population. However, both modernisation theory and the radical critique have been criticised for over-emphasising the economic dimensions of development. By the 1980s, 'development', in contrast, was seen as a multidimensional concept encapsulating widespread improvements in the social as well as the material well-being of all in society. In addition, it was recognised that there was no single model for achieving development and that investment in all sectors was required, including agriculture as well as industry. Above all, 'development' needed to be sustainable; it must encompass not only economic and social activities, but also those related to population, the use of natural resources and the resulting impacts on the environment. The multidimensional nature of the challenge of development is illustrated in Box A, which charts the major changes in thinking regarding approaches to issues of population growth.

The 1980s, however, have been referred to as the 'lost decade' in development. The suggestion is that, with the exception of significant examples of the East Asian 'Newly Industrialising Countries', the majority experience in the developing world during the 1980s was of 'development reversals':

> For many developing countries, there was a combination of declining international demand, increasing protectionism in the OECD, deteriorating terms of trade, negative capital flows, continuing high interest rates, and unfavourable lending conditions. The signs of progress up to the 1970s ground to a halt and in many cases went into reverse. Per capita national incomes in Latin America and Africa declined, investment declined (resulting in the deterioration of infrastructure and transport, communications, education and health care) and unemployment and underemployment grew.
>
> (Hewitt, 1992: 232–3)

World recession in the early part of the decade and the mounting 'debt crisis', particularly in the developing world, confirmed the limitations of past development strategies to promote and spread the benefits of economic growth. Figure 1.6 illustrates the persistent and generally mounting challenge throughout the 1980s of servicing debt in relation to export performance of regions of the developing world.

In the 1990s, after more than three decades of 'development', many developing countries have debt burdens which outweigh their Gross National Product several times over, as shown in Table 1.2. As seen, nine of the 'top ten' most indebted countries globally are in Africa. Income inequalities across the world have also increased rather than decreased with the passage of each United Nations Development Decade: in 1993, the richest 20 per cent of the world's population had a greater share of global income than they did in 1960, as shown in Figure 1.7. The ratio of richest to poorest within countries across the globe has increased from 30:1 to over 60:1 during the same period (UNDP, 1993).

Beyond such indicators of economic change into the late 1990s, Figure 1.8 presents insights into the achievements and continued deprivation in the developing world in terms of wider aspects of human development. It is evident that, as conceptions of development

## Box A

---

### *The relationship between development and population change*

Additions to the human race are currently running at over 90 million people a year. About 95 per cent of this growth is taking place in the developing countries of the world.

(Buckley, 1994: 1)

Of all the population growth that has taken place in the last 12,000 years, 75 per cent has occurred in the twentieth century, as shown in Table 1.1. In large measure, this growth has occurred in the developing world, as shown in Figure 1.5. By the early 1960s, world population was growing at 2 per cent per year, adding 70 million people annually, fundamentally through a steep decline in death rates without a similar drop in birth rates. At this time, the rapid rates of population growth in the developing world were considered to be a key constraint on economic development in these regions (and a cause of environmental degradation), the thinking being that the more people there were, the more resources they consumed and the greater the task of eliminating poverty, servicing the populations, creating employment and achieving development. 'Neo-Malthusian' ideas dominated the debate (which was largely hosted in the more developed regions). In line with Malthus' original work, it was thought that population growth could not be matched by food production and the consequences of this would be seen in terms of starvation and premature deaths. Curbing or controlling population growth was seen as a necessary condition if these countries were to become developed (and if the Earth was to be conserved, as seen in subsequent sections of this chapter).

**Table 1.1** *World population growth*

.....................................................

| Date | World population |
| --- | --- |
| 10,000 BC | 6 million |
| Birth of Christ | 250 million |
| 1000 AD | 350 million |
| 1750 | 800 million |
| 1830 | 1 billion |
| 1930 | 2 billion |
| 1960 | 3 billion |
| 1975 | 4 billion |
| 1987 | 5 billion |
| 1998 | 6 billion |

Source: Buckley (1994).

**Figure 1.5**  *Actual and projected world population growth, 1950–2020*

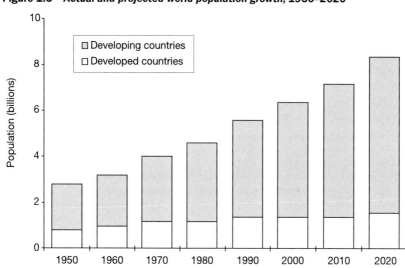

Source: Buckley, 1994.

This view dominated international thinking into the early 1970s and was a powerful force in promoting financial support for a whole host of programmes aimed at rapidly bringing down birth rates in the developing countries. High fertility in the underdeveloped nations was conceived to be a reflection of an unmet demand for contraception. The solution to development was therefore to provide family planning services within developing nations such that people could control their fertility and bring down growth rates. It was considered to be a simple technical exercise of taking Western contraceptive techniques and making them available in the developing world.

In practice, reducing fertility proved a complex and highly contested endeavour. In India, for example, it is accepted that public resentment to the mass sterilisation programmes in the 1970s was a major factor in the defeat of Indira Gandhi's government in 1978. In the ten months from April 1976 to January 1977, 7 million sterilisations were performed (Hanlon and Agarwal, 1977) but increased coercion, various malpractices and even forced sterilisation of men with more than three children in some states meant that the campaign was far from a success. The 'technical' issue in theory had become a highly political and ethical one in practice. Throughout the campaigns, however, India's government was supported by the World Bank, western aid agencies and western governments, both in monetary terms and through statements praising the government's 'hard-headedness' in the family planning drive.

At the First World Population Conference in 1974, there was much disagreement between the developed and developing countries over the relationship between population and development and the specific role of family planning programmes. A decade later at the Second World Population Conference, ideas of population 'control' were replaced by notions of population 'planning' based on women's reproductive rights and their freedom to plan the size of their family. In the 1980s, a more balanced view prevailed, suggesting that development itself (particularly in terms of maternal health and education) was likely to have the greatest impacts on reducing fertility. In short, it was acknowledged that the characteristics of underdevelopment, such as illiteracy, ill-health and lack of social services, were also those which were likely to impede fertility decline.

Into the 1990s, ideas concerning population and development now have women's equality and empowerment central within a much more holistic approach, 'which takes in poverty, women's status and the structure of society as well as fertility *per se*' (Johnson, 1995: 29). Women's issues dominated the Third World Population Conference in Cairo in 1994, where there was substantial debate, particularly regarding the Vatican's position on contraception and the fears of Islamic fundamentalists over the 'decadence' of sex education and women's reproductive rights. It is evident, however, that issues of population growth are now seen to be linked clearly to much wider issues of social change and development (many of which are expounded in subsequent chapters of this volume). In particular, there is an understanding of the need to recognise the rights of women as full human beings in society (not solely in the area of reproduction) and to meet women's needs not only for contraception, but for livelihood itself.

**Figure 1.6** *Debt service as a percentage of exports of goods and services, by world region*

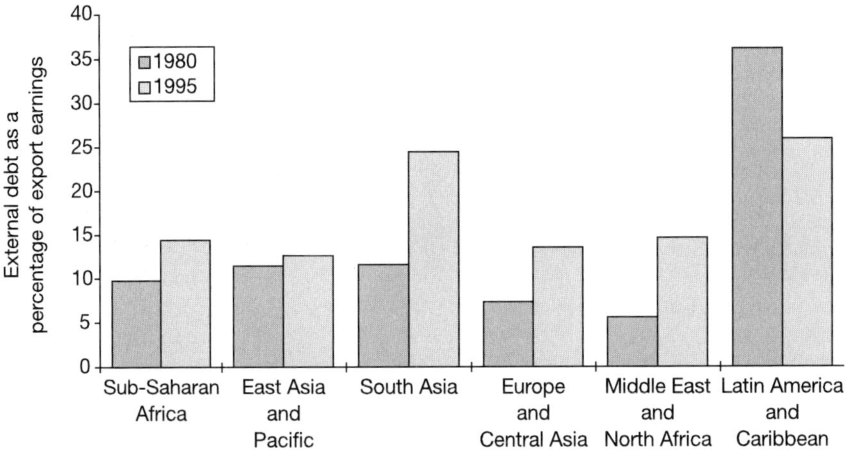

Source: World Bank (1997b).

have broadened to centre on the genuine needs of society, many people in the developing world in the late 1990s have extremely limited choices in their daily lives.

## Changing perceptions of the environment

The history of environmental concern is quite similar to that of development studies: although people have held and articulated varying attitudes towards nature stretching back many years, it is only since the 1960s that a coherent philosophy and language

**Table 1.2** *The world's 'top ten' most indebted nations, 1995*

·······························································

| Rank | Country | External debt as percentage of GNP |
|------|---------|------------------------------------|
| 1 | Nicaragua | 589.7 |
| 2 | Mozambique | 443.6 |
| 3 | Congo | 365.8 |
| 4 | Guinea Bissau | 353.7 |
| 5 | Angola | 274.9 |
| 6 | Côte D'Ivoire | 251.7 |
| 7 | Tanzania | 207.4 |
| 8 | Zambia | 191.3 |
| 9 | Malawi | 166.8 |
| 10 | Sierra Leone | 159.7 |

Source: World Bank (1997b).

surrounding the environment ('environmentalism' as defined by Pepper in 1984) can be identified.

In continuity with 'development thinking', it is possible to identify significant differences and changes over time concerning ideas about the environment; regarding society's relationship with nature; and in terms of the prescribed conservation requirements within modern environmentalism. Although the focus here is largely on 'mainstream' environmentalism as forwarded for example within successive conferences and publications of international institutions, the continued diversity within modern environmentalism should not be denied or underestimated. For example, there is a persistent and fundamental divergence between 'reformist/technocentric' and 'radical/ecocentric' environmentalism which is the source of much contemporary debate within sustainable development (Adams, 1996). Box B highlights the principal differences between these two philosophical standpoints on nature and society and the varied implications of each for conservation action.

**Figure 1.7** *The mounting divide between the rich and the poor in the world*

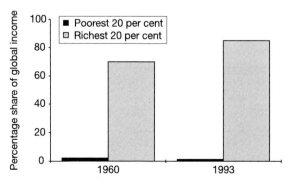

Source: UNDP (1996).

**Figure 1.8** *United Nations balance sheet of human development*

..................................................................................................................................

**Developing countries**

PROGRESS

DEPRIVATION

..................................................... HEALTH .....................................................

- In 1960–93 average life expectancy increased by more than a third. Life expectancy is now more than 70 years in 30 countries.
- Over the past three decades the population with access to safe water almost doubled – from 36% to nearly 70%.

- Around 17 million people die each year from curable infectious and parasitic diseases such as diarrhoea, malaria and tuberculosis.
- Of the world's 18 million HIV-infected people, more than 90% live in developing countries.

..................................................... EDUCATION .....................................................

- Between 1960 and 1991 net enrolment at the primary level increased by nearly two-thirds – from 48% to 77%.

- Millions of children are still out of school – 130 million at the primary level and 275 million at the secondary level.

..................................................... FOOD AND NUTRITION .....................................................

- Despite rapid population growth, food production per capita increased by about 20% in the past decade.

- Nearly 800 million people do not get enough food, and about 500 million people are chronically malnourished.

..................................................... INCOME AND POVERTY .....................................................

- During 1960–93 real per capita income in the developing world increased by an average 3.5% a year.

- Almost a third of the population – 1.3 billion people – lives in poverty.

..................................................... WOMEN .....................................................

- During the past two decades the combined primary and secondary enrolment ratio for girls increased from 38% to 78%.
- During the past two decades fertility rates declined by more than a third.

- At 384 per 100,000 live births, maternal mortality is still nearly 12 times as high as in OECD countries.
- Women hold only 10% of parliamentary seats

..................................................... CHILDREN .....................................................

- Between 1960 and 1993 the infant mortality rate fell by more than half – from 150 per thousand live births to 70.
- The extension of basic immunisation over the past two decades has saved the lives of about three million children a year.

- More than a third of children are malnourished.
- The under-five mortality rate, at 97 per thousand live births, is still nearly six times as high as in industrial countries.

..................................................... ENVIRONMENT .....................................................

- Developing countries' contribution to global emissions is still less than a quarter that of industrial countries, though their population is four times the industrial world's.

- About 200 million people are severely affected by desertification.
- Every year some 20 million hectares of tropical forests are grossly degraded or completely cleared.

..................................................... POLITICS AND CONFLICTS .....................................................

- Between two-thirds and three-quarters of the people in developing countries live under relatively pluralistic and democratic regimes.

- At the end of 1994 there were more than 11 million refugees in the developing world.

## Industrial countries

| PROGRESS | DEPRIVATION |
|---|---|
| **HEALTH** | |
| ● By 1992 life expectancy was more than 75 years in 24 or 25 industrial countries. | ● Nearly two million people are infected with HIV. |
| **EDUCATION** | |
| ● Between 1960 and 1990 the tertiary enrolment ratio more than doubled – from 15% to 40%. | ● More than a third of adults have less than an upper-secondary education. |
| **INCOME AND EMPLOYMENT** | |
| ● Between 1960 and 1993 real per capita GNP grew by more than 3% a year.<br>● The average annual rate of inflation during the 1980s was less than 5%. | ● The total unemployment rate is more than 8%, and the rate among youths nearly 15%. More than 30 million people are seeking work.<br>● The poorest 40% of households get only 18% of total income. |
| **WOMEN** | |
| ● Between 1970 and 1990 the number of female tertiary students per 100 male tertiary students studying science and technology more than doubled – from 25 to 67.<br>● Women now account for more than 40% of the labour force and about a quarter of administrators and managers. | ● The wage rate for women is still only two-thirds that for men.<br>● Women hold only 12% of parliamentary seats. |
| **SOCIAL SECURITY** | |
| ● Social security expenditures account for about 15% of GDP. | ● More than 100 million people live below the official poverty line, and more than 5 million are homeless. |
| **SOCIAL FABRIC** | |
| ● There are more than five library books and one radio for every person, one TV set for every two people. One person in three reads a newspaper. | ● Nearly 130,000 rapes are reported annually in the age group 15–59 |
| **ENVIRONMENT** | |
| ● Aggressive conservation measures and more appropriate pricing policies dramatically reduced energy use per $100 of GDP between 1965 and 1991 – from 166 kilograms of oil equivalent to 26 kilograms. | ● Each year damage to forests due to air pollution leads to economic losses of about $35 billion – equivalent to the GDP of Hungary.<br>● People in industrial countries consume nearly nine times as much commercial energy per capita as people in developing countries, though they constitute only a fifth of the world's population. |

Source: UNDP (1996).

## Box B

---

### *Modes of thought concerning humanity and nature*

It is argued that society's desire to manipulate nature, concomitant with an acceptance that the Earth nurtures our own existence, is inherent in the human condition. 'Technocentric' and 'ecocentric' refer to the two extreme positions. In reality, the distinction between these different perspectives is often blurred. As O'Riordan (1981) suggests, rarely is the world so neatly divided into two camps; rather we all tend to favour certain elements of both modes, depending on such factors as our changing economic status and the institutional setting or issue at hand. The categories should not, therefore, be thought of as rigidly fixed nor mutually exclusive.

| Environmental philosophies: | Technocentric | Ecocentric |
|---|---|---|
| | Human-centred: humanity has a desire to manipulate nature and make the world a more certain place in which to live. | Earth-centred: the Earth nurtures humanity's existence and should be treated with respect and humility. |
| **Green Labels:** | 'Dry Green' | 'Deep Green' |
| | Reformist in that the present economic system is accepted, but considered to require some gradual revision. | Radical in that quite rapid and fundamental changes in economy and society are desired. |
| | Belief in political status quo, but more responsible and accountable institutions. Self-regulation through 'enlightened conscience'. | Supports devolved, political structures with emphasis on self-reliant communities and pursuit of justice and redistribution across generations. |
| **Environmental management strategies** | Reliance on scientific credibility, modelling and prediction. | Management strategies geared to retaining global stability based on ecological principles of diversity and homeostasis. |
| | Promotes the appropriate manipulation of markets to create cost-effective solutions to environmental improvements. | New and fundamentally different conservation solutions required which are flexible and adaptable. |

Sustainable development through
rational use of resource, better
planning and clean technologies,
for example.

Alternative and appropriate
technologies.

Sources: compiled from Pepper (1996), O'Riordan (1981) and O'Riordan (1995).

In the 1960s, environmentalism was largely a movement reflecting European and American, white, middle-class concerns. The undesirable effects of industrial and economic development were beginning to be seen via a number of 'conspicuous pollution incidents' (Bartelmus, 1994: 5) and people were worried about the effects on their own lifestyles and health: 'after two centuries of industrialism and urbanisation, people now began to rediscover the idea that they were part of nature' (McCormick, 1995: 56). Environmentalists campaigned on issues such as air pollution and whaling and often received substantial support from the media. In contrast to earlier nature protection or conservation movements within these regions, environmentalism was overtly activist and political. The combination of actual changes in the environment and people's perceptions generally at this time brought widespread public support for the environmental movement, particularly amongst the younger groups. As Biswas and Biswas (1985) suggest, 'the environment and Vietnam became two of the major issues over which youth rebelled against the establishment' (p. 25).

For the new environmentalists, it was not solely their local outdoor environments which were perceived to be under threat, but human survival itself. A number of very influential 'global future studies' were published in the early 1970s which served to reinforce and spread the fears and influence of western environmentalists. For example, texts such as *The Population Bomb* (Ehrlich, 1968), *Blueprint for Survival* (Goldsmith *et al.*, 1972) and *The Limits to Growth* (Meadows *et al.*, 1972) modelled an ever-expanding population and mounting demands of society on a fundamentally finite resource base. In order to 'avoid the disastrous consequences of transgressing the physical limits of the Earth's resources' (Bartelmus, 1994: 5), urgent conservation actions (particularly population control in the developing world as seen in Box A) and 'zero-growth' in the world economy were required.

Not surprisingly, this environmental movement found little support in the developing nations. Many developing nations had only just gained independence and were sceptical regarding the motives behind proposals which seemed to limit their development objectives and remove sovereign control over resources. These underdeveloped nations saw their development problems as being linked to too little industry rather than too much and contrasted this with the position of the developed countries which used the bulk of resources and contributed most to the resulting industrial pollution. Similarly, it was hard for representatives of the developing world to consider how their relatively impoverished citizens could compete with affluent consumers of the West in terms of responsibility for the depletion of resources.

Development and conservation at this time were portrayed as incompatible: resources were thought to be finite, and pollution and environmental deterioration were considered the inevitable consequences of industrial development. In conceiving the 'environment' as the stocks of substances found in nature, by definition these resources were ultimately considered to be limited in quantity. In turn, therefore, the global future predictions such as *The Limits to Growth* gave little attention to the social, technological or institutional factors which affect the relationship between people and resources (Biswas and Biswas, 1985). Further to such environmental determinism, these reports were ahistorical in the sense that they gave no attention to how or why the world is divided into rich or poor, for

**Plate 1.2** *The 'inevitable' consequences of development? Industrial air pollution*

Source: Gordon Walker, Staffordshire University.

example. They were also apolitical in considering the future of the Earth as the overriding and paramount concern, with no consideration of how the solutions advocated would favour some nations or groups over others.

It was not until the mid-1970s that the fears of the developing countries were overcome and changing ideas of the environment ensured a greater participation of these nations in the environmental debate. In 1971, the UN hosted a seminar on Environment and Development at Founex, Switzerland. There were two very important outcomes of this meeting. First, there was a much more enlightened appreciation on the part of the international community of the position of the developing nations: their fears over the economic effects of environmental protection policies, their desire for industrialisation as well as wider social and cultural development, and the nature of their own environmental problems. Second, the conception that environment and development problems were incompatible was overthrown as ideas of the environment were expanded from solely Western concerns to include those which stem from a lack of development.

By the time of the UN Conference on the Human Environment in 1972, the environmental movement had 'come of age' and environmental issues were clearly on the international political agenda, as evidenced by the participation in Stockholm of 113 countries. Adams (1996) refers to the 1972 meeting as a 'watershed in the emergence of sustainable development' (p. 358). Although the primary impetus for the conference had been the developed world's concerns about the effects of industrialisation, the dialogue between government representatives, and within parallel meetings of non-governmental organisations (NGOs), soon moved to wider issues including the relationship between environment and development issues. The term 'pollution of poverty' was used for the first time at Stockholm to refer to the environmental concerns of the poor, such as lack of clean water or sanitation, which threatened life itself for many in the developing world. It also encompassed the emerging recognition that a lack of development could also cause environmental degradation, a concept which is now shaping much thinking and action towards sustainability, as detailed in subsequent chapters.

In the year following the Stockholm conference, the United Nations Environment Programme (UNEP) was established. The intention was that UNEP would serve as an 'international environmental watchdog'

**Figure 1.9**   *The World Conservation Strategy objectives of conservation*

......................................................................................................

1 **The maintenance of essential ecological processes and life-support systems** such as soil, forests, agricultural systems, and coastal and freshwater systems. This meant managing cropland, protecting watersheds and coastal fisheries, and controlling the discharge of pollutants.

2 **The preservation of genetic diversity** for breeding projects in agriculture, forestry and fisheries. This meant preventing the extinction of species, preserving as many varieties as possible of crop and forage plants, timber trees, animals for aqua-culture, microbes and other domesticated organisms and their wild relatives, protecting the wild relatives of economically valuable and other useful species and their habitats, fitting the needs of ecosystems to the size, distribution and management of protected areas, and co-ordinating national and international protected area programmes.

3 **Ensuring the sustainable use of species and ecosystems**. This meant ensuring use did not exceed the productive capacity of exploited species, reducing excessive yields to sustainable levels, reducing incidental take, maintaining the habitats of exploited species, carefully allocating timber concessions and limiting firewood consumption, and regulating the stocking of grazing lands.

......................................................................................................

Source: McCormick (1995).

in terms of being responsible for the monitoring of global environmental change. Although its monitoring role has been limited (Werksman, 1996), UNEP remains the UN's primary environmental policy co-ordinating body and its role has been strengthened in subsequent international fora (see Chapter 3).

In summary, by the late 1970s, important changes in thinking regarding both the environment and development were causing the two previously separate issues to be seen as interdependent concerns. As a result, the interdependence of the developed and developing worlds was also recognised. The challenge for the 1980s was to formulate policies for action which would integrate the environment and development in practice. In 1980, the World Conservation Strategy (WCS) was published by the International Union for the Conservation of Nature and Natural Resources (IUCN), in conjunction with the United Nations Environment Programme and the World Wildlife Fund (now the World Wide Fund for Nature). It has been referred to as the 'launchpad' for the concept of sustainable development (Mather and Chapman, 1995: 248).

Within the WCS, for the first time, development was suggested as a major means of achieving conservation rather than as an obstruction to it: 'human benefits would follow from appropriate forms of environmental management' (ibid.: 247). The WCS identified three objectives for conservation, as shown in Figure 1.9. Conservation itself was defined as being more than nature preservation (which had been the traditional focus of IUCN since its formation in 1952) and the strategy encompassed many more positive ideas concerning sustainable utilisation and enhancement of the natural environment.

In the early 1980s, the World Commission on Environment and Development was charged explicitly with formulating proposals for dealing with the critical environment and development issues of the globe. The 'Brundtland Report' (named after its Chair, the former Prime Minister of Norway, Gro Harlem Brundtland) extended the ideas of sustainable development significantly beyond those of the WCS and did much to disseminate the popular and political use of the term. By 1992, for example, the report had been translated into twenty-four languages (Finger, 1994). Principally, the WCED gave greater attention to development concerns: to the challenge of overcoming poverty and meeting basic needs, and to integrating the environment into economic decision-making. For this, reviving economic growth was considered essential. No longer was it a question of whether development was desirable; economic growth was central to the Commission's proposals for environmental protection

**Plate 1.3**  *The pollution of poverty: hazardous housing on a Calcutta roadside*

Source: author.

and 'degraded and deteriorating environments were seen to be inimical to continued development' (Mather and Chapman, 1995: 248). But new forms of economic growth would be the key to sustainable development; growth must be less energy intensive and more equitably shared, for example.

For some, the continued prominence given in the report to economic growth in future development has suggested a 'comfortable reformism'. Although the substantial political and economic changes required in future to achieve sustainable development were identified by the WCED (see Figure 1.2), it has also been suggested that it did not go far enough in terms of identifying the barriers or obstacles to change in these sectors (Starke, 1990).

In 1988, the Centre for Our Common Future was established to spread the message of the Brundtland Report and the ideas of sustainable development. When it was announced that there was to be a UN Conference on Environment and Development (UNCED) in 1992, the Centre became primarily concerned with mobilising intergovernmental organisations, governments and NGOs into the UNCED process. In subsequent chapters, the specific agreements reached at Rio de Janeiro and the impacts on the actions of varied institutions of development at different levels are discussed.

## The 1990s: sustainability, globalisation and debt

In the late 1990s, environmental issues are firmly established in development thinking. By this time, it has been seen that the concept of sustainable development encapsulates notions of development based in the reality of local environments and the needs of the poorest sectors in society which are far removed from the unilinear, econometric models of the early 1960s, for example. Insight into the multidimensional nature of this development challenge in the future has also come through expanded notions of the environment and the functions it plays in human societal development. Critically, it is recognised that sustainable development is a *global* challenge; ultimately, the achievement of environment and development ends in any single location or for any group of people is connected in some way to what is happening elsewhere, for others.

This recognition of the threat of society to the life-support systems of the Earth is argued to be just one feature of the 'new times' in which we live (Johnston *et al.*, 1995). At the end of the twentieth century,

## Box C

## *The unevenness of globalisation*

Despite the global character of many major processes of economic, political, environmental and social change in the world at the turn of the twenty-first century, it should not be taken that globalisation affects all people or all areas of the globe equally. It is evident that some parts of the world are 'left out' in the sense that they are not part of a network of communications or do not receive multi-national investment, for example. A simple illustration is that half of the world's population has never made a telephone call (Potter *et al.*, 1999). Table 1.3 highlights the global inequalities in the share of Internet hosts, the 'leading edge of the globalisation of culture' as suggested by Knox and Marston (1998). By 1996, there were nearly 10 million computer systems connected to the Internet, 64 per cent of which were in the United States.

**Table 1.3   *Access to Internet hosts, selected countries, 1996***

| Country | Percentage of Internet hosts | Percentage of world population |
| --- | --- | --- |
| United States | 63.7 | 4.82 |
| United Kingdom | 4.8 | 1.0 |
| Canada | 3.9 | 0.5 |
| Australia | 3.2 | 0.32 |
| Japan | 2.8 | 2.3 |
| Finland | 2.2 | 0.09 |
| Brazil | 0.21 | 2.83 |
| China | 0.02 | 21.9 |
| India | 0.008 | 17.0 |

Source: Knox and Marston (1998).

Foreign direct investment (FDI), made by private companies overseas, is the major driving force of economic globalisation. A good part of such investments abroad are concentrated in the hands of large trans-national corporations. When these large companies are classified by home country, as in Table 1.4, it is seen that the major investors overseas into the late 1990s have been from the US, UK and Japan. Hong Kong is also an emerging source of foreign direct investment. Table 1.5 shows where such investments go. In large measure, it is seen that the major investors overseas are investing in each other. Although the share of inward investment to developing countries rose slightly in the early 1990s, it continues to go to a very limited number of countries (UNCTAD, 1997). For example, the bulk of recent investment in Asia has been in China specifically (which was the largest developing country recipient in 1996). Brazil now has the largest share of inward investment in Latin America as a whole. In contrast, the African continent has a small and declining relative share of multi-national investment and as such could be suggested to be only loosely connected or even excluded from the world economy.

**Table 1.4** *Stocks of outward foreign direct investment (FDI) by home country as a percentage of total FDI, 1980–96*

| Country | 1980 | 1990 | 1996 |
|---|---|---|---|
| United States | 40 | 25 | 24 |
| United Kingdom | 15 | 15 | 15 |
| Japan | 7 | 14 | 7 |
| Germany | 8 | 9 | 8 |
| Hong Kong | – | 1.4 | 8 |

**Table 1.5** *Stocks of inward FDI by major host country/region as a percentage of the total, 1980–96*

| Country/region | 1980 | 1990 | 1996 |
|---|---|---|---|
| Western Europe | 42 | 44 | 30 |
| United Kingdom | 13 | 13 | 9 |
| United States | 17 | 24 | 24 |
| Japan | 1 | 1 | <1 |
| Africa | 3 | 2 | 1 |
| Asia | 8 | 10 | 24 |
| Latin America and Caribbean | 13 | 7 | 11 |

Sources: Allen and Hamnett (1995a), UNCTAD (1997).

In summary, rather than globalisation leading to common experiences or some shared path to development, processes of globalisation 'work themselves out unevenly and in turn, are shaped by the patterns of uneven development previously laid down' (Allen and Hamnett, 1995a: 234).

people across the world are experiencing not only unprecedented rates and degrees of environmental change, but simultaneously also huge economic, political, social and cultural change. Furthermore, and as seen in the analysis of changes in development thinking and environmentalism, there is substantial uncertainty concerning present conditions and future directions in all these arenas. It appears that 'recent past experiences are being less and less a sound basis upon which to plan our actions' (ibid.: 4).

'Globalisation' is the term widely used to encompass these various processes of change through which interactions between different regions are increasing and the world becomes ever more global in

**Table 1.6** *The state and corporate power,*
*1994*

| Country or *corporation* | Total GDP or corporate sales (US$ billions) |
|---|---|
| Indonesia | 174.6 |
| *General Motors* | 168.8 |
| Turkey | 149.8 |
| *Ford* | 137.1 |
| South Africa | 123.3 |
| *Toyota* | 111.1 |
| *Exxon* | 110.0 |
| *Shell* | 109.8 |
| Norway | 109.6 |
| Poland | 92.8 |
| *IBM* | 72.0 |
| Malaysia | 68.5 |
| Venezuela | 59.0 |
| Pakistan | 57.1 |
| *Unilever* | 49.7 |
| Egypt | 43.9 |
| Nigeria | 30.4 |

Source: *New Internationalist* (1997b).

character. Whilst global links and interconnections between places and peoples around the world have existed previously (through colonial ties, for example), the nature, extent and depth of contemporary processes of globalisation are new. As Allen and Hamnett (1995b) suggest, it is not the global scope of movements of people or resources currently, but the immediacy and intensity with which we can now experience other parts of the globe, which is unprecedented and characteristic of globalisation at the end of the 1990s. However, whilst the world is becoming more global, it does not necessarily mean that it is becoming more uniform, as discussed in Box C.

Economic processes of globalisation are controlled substantially by multi-national enterprises, firms which have operations in more than one country with a home base in their country of origin. Their influence comes partly from their size: many multi-national companies have larger sales and income than whole countries, as illustrated in Table 1.6. Such large corporations can now subcontract production, research and development facilities to branch plants located at great distances from the centres of demand. Furthermore, the 1990s have seen the emergence of 'trans-national corporations', new forms of organisation which have developed to ensure a market presence in each of the major economic regions of the world and which may no longer have a headquarters within a particular country. Rather, all economic activities are integrated at the global scale and all functions located wherever it is most favourable economically; 'this may involve a geographic separation of functions to take advantage of what different places have to offer or an overlapping set of activities wherever they have a market presence' (Allen, 1995: 61).

The most dramatic increase in economic globalisation, however, has taken place, not in production, but in financial speculation. For

**Plate 1.4** *Multinational presence in daily lives world-wide: Muslim prayers in Kuala Lumpur*

Source: Popperfoto.

example, developments in communication technologies now enable 24-hour global trading on the world currency markets with money being switched around the globe at high speeds. Money can be placed virtually instantaneously wherever it produces the highest returns (and shifted subsequently to other locations to secure greater gains). Such financial speculation has little to do with the 'real economy', in providing capital for manufacturing, agriculture or service industries, for example. It can, however, disrupt the plans of national governments, such as those to control inflation or the movement of exchange rates, with severe short- and long-term development impacts, as was seen in the crises of many South-East Asian economies (the 'newly industrialising countries' in the 1980s) from mid-1997.

Sustainable development is also being pursued in the 1990s in the context of a continued reduction in world economic growth rates. Historically, such economic pressures have been important factors behind significant political changes within and between states, and this has again been demonstrated in recent years. The currency crisis in 1998 in Indonesia, for example, led directly to the removal of President Suharto after more than thirty years in power. Economic and political changes were also very much interwoven in the 1989 and 1991 revolutions in Eastern Europe and the USSR respectively. Less markedly, but almost universally, there has been a redefinition of the role of the state in the 1990s in response to the slowdown in economic growth. Typically, the move has been towards a less interventionist state and a greater role given to the market in income generation and wealth distribution.

In the late 1990s, for many nations in the developing world, their entry into the world economy is increasingly defined by the neo-liberal policies of the World Bank (WB) and the International Monetary Fund (IMF). As already suggested, many developing countries began to experience severe balance of payments difficulties in the early 1980s. In the previous decade, debts on capital borrowing by developing countries from private commercial banks had increased by 20 per cent per year (World Bank, 1983). However, problems of servicing these debts emerged during the 1980s as recession hit in the industrialised countries. Interest rates on borrowing rose, demand for exports from those borrowing countries declined and the prices received, particularly of commodity products, fell. Whilst the commercial banks became increasingly worried about their bad debts, the insolvency of particularly the middle-income debtor countries threatened the international financial system as a whole. As such, debt became the concern of the two 'mainstays of the global economic order', the WB and the IMF. The assessment in the early 1980s was that the economic crisis in developing countries was more than a temporary liquidity issue (as it had been conceived in the 1970s). Rather, comprehensive, longer-term solutions were required, based on packages of broad policy reforms in indebted nations.

The term structural adjustment programme (SAP) is used to refer to the generic activities of the IMF and WB in the arena of policy reform. The central objective of SAPs as defined by the World Bank is to 'modify the structure of an economy so that it can maintain both its growth rate and the viability of its balance of payments in the medium term' (Reed, 1996: 41). The first SAP was implemented in Turkey in 1980 and by the end of the decade 187 SAPs had been negotiated for 64 developing countries (Dickenson *et al.*, 1996: 265). Although each package is tailored for the particular country, SAPs generally include many or all of the elements listed in Figure 1.10. It has been argued that the impacts of SAPs have now gone far beyond the original

**Figure 1.10** *The principal instruments of structural adjustment*

..................................................

Cuts in:
- government expenditure
- public sector employment
- real wages

Pricing policies designed to:
- eliminate food subsidies
- raise agricultural prices
- cost recovery in public services

Trade liberalisation involving:
- currency devaluation
- credit reform
- privatisation of state-owned institutions
- higher interest rates

..................................................

national contexts for which they were designed, to become an instrument for global economic restructuring (Reed, 1996) and through the conditions attached, they enable the IMF and WB to 'virtually control the economies' of many developing nations (Hildyard, 1994: 26). Certainly, these international institutions currently influence development policy and planning in the developing world to an unprecedented extent and are important actors in determining the prospects for sustainable development in the future.

## Conclusion

Evidently, the idea of sustainable development is not new but has a substantial history. Similarly, the contemporary debate as to how best to secure both development and conservation ends in the near future has also been seen to have long-standing origins. However, what was new in the 1980s was the way in which the two literatures of development and environmentalism came closer together, recognising the significant and interdependent nature of these goals. By the end of the 1990s, the widespread suggestion is that the world itself is characterised by unprecedented rates and degrees of economic, political and social change. This is the context in which the challenges and opportunities of sustainable development actions are being debated, designed and carried out.

## Discussion questions

* What are the principal lessons for the international community of nearly four decades of 'development'?
* Trace the major ways in which debt could compromise a country's ability to foster sustainable development policies.
* Consider the ways in which global deforestation can be viewed as both a major environment and development problem.

# Further reading

Adams, W.M. (1990) *Green Development*, Routledge, London.

Allen, J. and Hamnett, C. (eds) (1995b) *A Shrinking World? Global Unevenness and Inequality*, The Shape of the World: Explorations in Human Geography series, no. 2, Oxford University Press, Oxford.

Johnston, R.J., Taylor, P.J. and Watts, M.J. (eds) (1995) *Geographies of Global Change*, Blackwell, Oxford.

McCormick, J. (1995) *The Global Environment Movement*, 2nd edn, Wiley, London.

Potter, R.B., Binns, J.A., Elliott, Jennifer A. and Smith D. (1999) *Geographies of Development*, Addison Wesley Longman, Harlow.

World Commission on Environment and Development (1987) *Our Common Future*, Oxford University Press, Oxford.

# The challenges of sustainable development

- How development processes depend on the natural environment
- Past failings in development and environmental conservation
- The pollution of poverty
- Gender and the environment
- The global challenges of personal and political change

## Introduction

Sustainable development is fundamentally about reconciling development and the environmental resources on which society depends. This chapter identifies some of the major characteristics of past development patterns and processes, and highlights the core impacts of these on the various resource functions which the environment provides. This is essential for understanding the challenges of sustainable development for decision-makers at all levels, in terms of supporting agricultural and industrial production in the future, dissipating the by-products of production and consumption, 'cleaning up' the effects of past development and protecting the environment from further damage.

## The centrality of resources in future development world-wide

All forms of economic and social activity make demands on the resource base: as raw materials such as soil and water within agricultural production, as sources of inputs and energy into industrial production or in the construction and maintenance of human settlements and urban lifestyles, for example. Whilst absolute resource scarcities (as predicted in several of the global future scenarios discussed in Chapter 1) have not generally materialised, economic development in the past has been closely correlated with mounting rates of resource extraction. This is seen in the case of

global water withdrawals which have expanded during the twentieth
century at rates in excess of population growth, as shown in Figures
2.1 and 2.2. For many developing and more developed nations,
accessing quality water supplies in the future will depend on the
resolution of complex political and institutional challenges associated
with harnessing the sources of river and lake systems that extend
across more than one national boundary. Indeed, many similarities
have been noted between the 'impending water crisis' (Biswas, 1993)
and the issues of supply and demand of oil in the 1970s and early
1980s, particularly in terms of the implications for political and
economic stability in regions such as the Middle East.

**Figure 2.1**   *Total global water use, 1940–87*

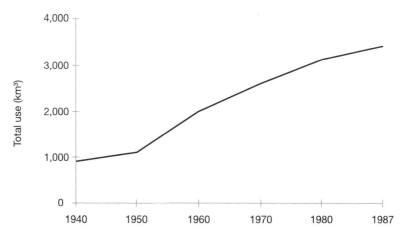

**Figure 2.2**   *Per capita global water use, 1940–87*

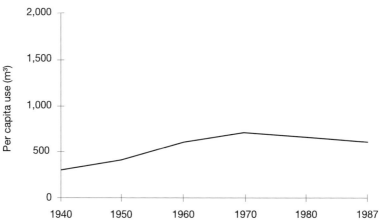

Note: Data is not yet available for the 1990s.

Source: Mather and Chapman, 1995.

All production and consumption activities also produce wastes in the form of various gases, particulate matter and chemicals. Where the rate of waste generation exceeds the natural capacity of the atmosphere, oceans, vegetation or soils to absorb these, detrimental effects occur on human health and on the operation of ecological systems. In 1991, the first deaths from air pollution in Britain for more than thirty years occurred in London, where smog levels due to traffic fumes built up over four particularly windless days in December and 160 people died as a result (Bown, 1994). There is also the ongoing challenge of finding the space to dispose of solid wastes, including those produced through domestic consumption. The volume of municipal wastes in industrialised countries is growing every year (Redclift, 1997). For example, the inhabitants of New York City alone throw away approximately 26,000 tons of solid waste every day. This is taken to a landfill site on Staten Island which,

> at the end of its life in 10–20 years, will consist of four pyramids of up to 435 feet high and become the highest point on the coast between Maine and the tip of Florida. When all the space is used up, the inhabitants of New York will have nowhere to put three quarters of their rubbish.
>
> (Gourlay, 1995: 6)

The challenges of sustainable development also include 'cleaning up' the pollution impacts of past development. Often pollution impacts do not occur immediately, but may take many years to build up or be recognised. Box D highlights the unexpected problem of arsenic contamination of water caused by development initiatives which had been aimed very directly at improving the health status of people in Bangladesh. Providing alternative water supplies to an estimated 43,000 affected villages spread throughout the country is a substantial challenge, not least because of the powerful government and aid interests tied to the continued sinking of tubewells (Pearce, 1998a; 1998b). In the case of many countries of Eastern Europe and the former Soviet Union, it has only been since the collapse of the communist regimes between 1988 and 1991 that previously restricted information has brought to light the substantial pollution legacy of past development. The financial challenge alone is considerable:

> According to the German Institute of Ecological Economic Research, bringing eastern Germany's environment up to the new Germany's standards will be a 10 year, $249- to $308-billion effort. Poland's environment minister estimates that improving conditions in his country will cost $20 billion over the next ten to twenty years. Czechoslovakians (*sic*) believe they will need to spend at least $23.7 billion on

## Box D

---

### *The unexpected environmental impacts of development*

In the early 1970s, multi-lateral and bi-lateral donors assisted Bangladesh in sinking thousands of tubewells to provide clean water drawn from the sands and silt of the Ganges floodplain. Currently, an estimated 95 per cent of drinking water in Bangladesh is provided through over 3 million tubewells. It has recently been found, however, that people using many of these wells are, in fact, being slowly poisoned by naturally occurring arsenic in the alluvial sediments of the delta, which appear to be released by chemical changes prompted by the fluctuations in water levels caused by pumping.

It is estimated by Professor Chakraborti of the University of Calcutta, who first uncovered the problem, that as many as 30 million people in Bangladesh may be drinking contaminated water. A fifth of the nation's drinking water is considered to contain up to one hundred times the maximum level of arsenic recommended by the World Health Organisation. A real problem in determining the numbers of people affected is that the early symptoms of increased skin pigmentation take ten years to develop. Many of those who develop these skin conditions, however, go on to develop internal cancers.

There is no effective treatment for arsenic poisoning: poisonous wells need to be identified and alternative sources of water found. Testing is relatively simple, but with approximately one tubewell for every four rural households, it is a huge task. There remains much uncertainty in the scientific and development communities as to what the response should be, including the ethics of stopping further tubewells, switching to surface sources (and risking other negative health impacts) and the viability of alternative sources such as rainwater harvesting.

Sources: Pearce (1998a; 1998b).

pollution control in the next 15 years. According to one estimate, the Soviet Union must spend 100 billion rubles ($11.75 billion) immediately and 10 billion rubles annually just to reduce air pollution to the accepted limits.

(French, 1990: 40)

The global nature of the challenges of sustainable development in future is clearly illustrated within the issue of biodiversity. The potential value of diversity within ecosystems, species and genetic materials is impossible to quantify. However, it is considered that the current rate of species extinction is exceeding that at which new species are emerging by over a million times (Mather and Chapman, 1995: 120). The costs of such losses will fall to humankind as a whole, through the disruption of the complex food chains and webs which constitute the ecologies of the planet and in terms of undiscovered opportunities of gene use within medicines, for example. Figure 2.3

**Figure 2.3** *'Megadiverse' countries of the world*

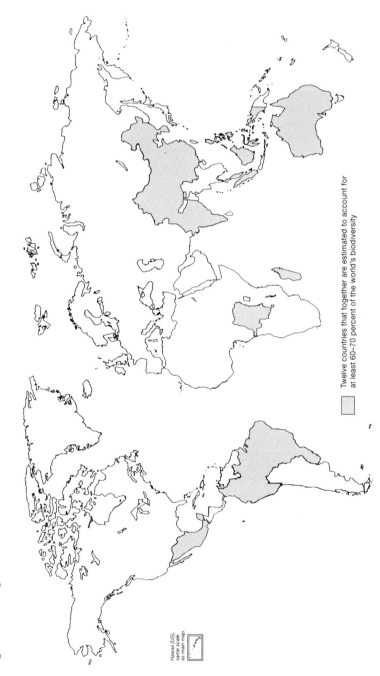

Hawaii (US),
same scale
as main map.

Twelve countries that together are estimated to account for
at least 60–70 percent of the world's biodiversity

Source: World Bank, 1992

identifies the twelve countries around the globe which together are estimated to account for the majority of the world's biodiversity. Evidently, these are largely the areas of the world's remaining tropical forests, which are particularly species diverse. Yet past processes of development, including the expansion of agricultural areas and commercial logging, have led to an acceleration of tropical deforestation during the twentieth century (Potter *et al.*, 1999).

## What has been wrong with past development?

In Chapter 1 it was seen how ideas about how best to achieve development and as regards society's relationship with the environment changed over time, in part as evidence emerged 'on the ground' about the successes and failures therein. The following sections provide greater detail on the limitations of past patterns and processes of development across the globe which have underpinned the call for sustainable development. In so doing, greater insight is gained into what sustainable development means in practice: the nature and extent of the challenges for action, for whom and where, for example.

## Inequalities in access to resources

The influence of issues of population in shaping modern environmentalism was seen in Chapter 1. Currently, it is not doubted that in some countries the ability of governments to provide basic needs of shelter, food, water and employment for their populations becomes increasingly difficult with rising numbers. For example, between 1980 and 1995, the population of Kenya increased every year by more than three-quarters of a million people (an average of 3 per cent per year). In contrast, the population of Spain (a country of similar geographic area) grew by only 2 million people over the total fifteen-year period (World Bank, 1997b). However, it is now more generally appreciated that inequalities in people's access to resources and the resultant ways in which they use them constitute greater challenge for sustainable development than issues of population numbers *per se* (WCED, 1987).

As will be seen throughout this text, there is tremendous diversity in terms of people's access to resources of all kinds, from land to investments in health care, for example. The case of commercial energy sources illustrates how inequalities in access to resources can be considered at a number of levels. Whilst energy can be derived from various sources (see Soussan, 1990), a conventional distinction is made between commercial fuels, such as oil, coal, gas and electricity, which have a commercial value and are often traded between countries, and non-commercial fuels, such as wood, charcoal and plant/animal residues, which are less widely bought or sold, certainly within formal markets. Figure 2.4 illustrates that the consumption of commercial energy world-wide is concentrated in the regions of Europe, North and Central America, and Asia. The disparity in energy use between regions is shown, for example, in the case of Africa, which has 13 per cent of the world's population, but less than 3 per cent of the total energy use.

It has been estimated that the annual commute by car into New York City alone uses more oil than the whole of Africa (excluding South Africa) in one year (Edge and Tovey, 1995: 319). In Africa, traditional fuels accounted for an average of 35 per cent of total consumption in 1993 (World Resources Institute, 1996). However, this continental scale of analysis (which includes major oil-producing countries such as Libya) masks some very high levels of dependence on biomass sources of energy for particular countries, such as Ethiopia (90 per cent), Malawi (92 per cent), Sierra Leone (83 per cent) and Tanzania (92 per cent).

Tables 2.1–2.3 emphasise further inequalities in access to the basic opportunities for development. In Table 2.1, it is seen that generally the wealthier countries have delivered basic environmental improvements in terms of access to clean water supplies for the

**Figure 2.4** *Share of world population and primary energy consumption*

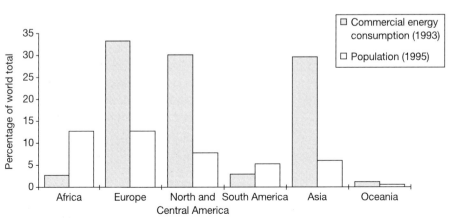

Source: World Resources Institute, 1996.

**Table 2.1** *International gaps in access to safe water supplies*

| World Bank country classification based on GNP per capita | Country | Percentage of total population with access to safe water supplies |
|---|---|---|
| Low Income | Mozambique | 28 |
| | Bangladesh | 83 |
| | Nigeria | 43 |
| | India | 63 |
| | Zimbabwe | 74 |
| | Sri Lanka | 57 |
| Middle Income | Ecuador | 70 |
| | Jamaica | 70 |
| | Malaysia | 90 |
| | Namibia | 57 |
| High Income | UK | 100 |
| | Australia | 95 |
| | USA | 90 |
| | Norway | 100 |

Source: World Bank (1997b).

**Table 2.2** *Rural–urban gaps in access to safe water supplies (percentage of population with access to services)*

| Country | Rural | Urban |
|---|---|---|
| Mozambique | 40 | 17 |
| Bangladesh | 97 | 99 |
| Zimbabwe | 64 | 99 |
| Namibia | 42 | 87 |
| India | 79 | 85 |
| Sri Lanka | 49 | 87 |
| Ecuador | 38 | 56 |

Source: UNDP (1996).

majority of their populations. Table 2.2 also shows how regional differences within a country can occur such as between urban and rural people. Here the pattern is less straightforward: urban populations are not necessarily better served, as shown by the case of Mozambique. Inequalities in access can also exist within a community, such as seen in Table 2.3, between socio-economic groups in a city, such as in Accra, where wealthier suburbs fare better than poorer on a number of environmental dimensions. However, it will also be seen in subsequent sections of this chapter that wealth is not the only factor determining access to resources and opportunities for development. Differences

Table 2.3 *Access to basic water services in poor, middle-class and wealthy neighbourhoods of Accra, Ghana, 1991–2 (percentage of sample households)*

| | Poor | Middle class | Wealthy |
|---|---|---|---|
| No water at source of residence | 55 | 15 | 4 |
| Share toilets with more than ten households | 60 | 17 | 2 |

Source: World Resources Institute (1996).

may also exist within the household; women, for example, regularly 'own a very small proportion of the natural resources and often face discrimination, when compared with men, in obtaining land, education, employment and housing' (Satterthwaite *et al.*, 1996: 16).

Inequality in access to resources threatens the prospects for sustainable development in many ways. Primarily, it confines large numbers of people to poverty which often leaves them with no choice but to degrade and destroy the resource base on which their future livelihood depends. It is also such inequalities which allow a minority of people globally, within each nation and even at the community level, to use resources in a wasteful manner or in ways which cause environmental damage. The call for sustainable development in the future stems from the fact that such inequalities not only are morally wrong but also threaten the environmental basis for livelihoods and development aspirations across the globe.

## The 'geographical retreat' of poverty

In 1990, the World Bank (using 1985 data) estimated that just over 30 per cent of the population of the developing world, more than one billion people, lived below the poverty line, identified as 370 'purchasing power parity' dollars (see Table 2.4). At that time, it was predicted that global poverty would decline into the 1990s. However, Todaro (1997), using World Bank data, calculated that, by 1990, the number of people living in poverty had actually increased by a further 82 million people. Once again, this scale of analysis masks substantial variations within these developing regions, particularly the large proportions of people in poverty in a number of heavily populated countries such as Bangladesh, which had 78 per cent of its population in poverty in 1990 (Todaro, 1997). Furthermore, high per capita incomes *per se*, on any scale, do not guarantee the absence of significant numbers of absolute poor: 'absolute poverty can and does

**Table 2.4** *The extent of poverty in the developing world in 1985*

| Region | Number below the poverty line (millions) | Percentage of population |
|---|---|---|
| Sub-Saharan Africa | 184 | 47.6 |
| East Asia | 182 | 13.2 |
| South Asia | 532 | 51.8 |
| Middle East and North Africa | 60 | 30.6 |
| Latin America and the Caribbean | 87 | 22.4 |
| All developing countries | 1,051 | 30.5 |

Source: Todaro (1997) based on World Bank (1990a) data.

exist as readily in New York City as it does in Calcutta, Cairo, Lagos or Bogota, although its magnitude is likely to be much lower in terms of numbers or percentages of the total population' (Todaro, 1997: 151).

It should be noted that 'poverty' itself is a contested term, open to many different definitions and interpretations. Whilst the World Bank's definition prioritises income or levels of private consumption, other institutions and authors incorporate a number of components of 'human development' and deprivation. For example, since 1990, the United Nations Development Programme has developed a 'Human Development Index' encompassing measures of real purchasing power, education and health. Further conceptions of poverty now regularly include factors of security, autonomy and self-esteem, and are closely aligned with the new thinking on 'social exclusion' in both developing and developed world contexts (for a discussion, see IDS Bulletin, 1998).

Concepts of poverty in the developing world now also include environmental dimensions: the command over resources which people have (through their ownership or via membership of particular social groups), coupled with their capacity to withstand environmental stresses and shocks, and including the ability people have to make effective and sustainable use of these 'entitlements' (Leach and Mearns, 1991). Poor people are regularly portrayed as both the 'victims and unwilling agents' of environmental degradation in developing regions (ibid.). Their environmental concerns are those associated with immediate survival needs, such as for fuel, access to clean water and sanitation or in securing productive lands. Their

Plate 2.1  *Poor people in poor environments*

Source: Hamish Main, Staffordshire University.

poverty restricts the options they have in terms of resource management: they may have to cultivate marginal lands, live in unsafe housing or remove remaining woodlands in order to survive in the short term, often with detrimental effects on the resource base and their own longer-term livelihoods. Figure 2.5 illustrates some of these circular linkages between poverty and the environment. It shows how poverty can contribute to environmental stress, such as through the cultivation of steep hillsides, but also how the adverse ecological effects associated with people endeavouring to secure their basic needs in turn threaten the health and survival of those same people, such as through the aggravation of sedimentation and flooding.

World-wide, the poor and most vulnerable groups often live in the environmentally most fragile areas, both rural and urban. The poor 'retreat' into certain geographical areas, because those environments are poor, where, fundamentally, demand for other uses is low. So, for example, the poor are increasingly concentrated in areas where the characteristics of soils demand high levels of investment in order to become productive, on lands of low commercial value where they are less likely to be evicted (such as alongside hazardous installations or railway tracks), or in 'frontier' areas such as forest edges, where infrastructure and other services are undeveloped. In the mid-1980s, it

**Figure 2.5** *The poverty and environment connection*

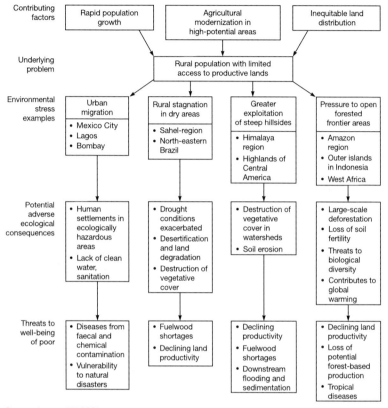

Source: Leonard (1989).

was suggested that 57 per cent of the rural poor and 76 per cent of the urban poor were resident in areas where ecological destruction and/or severe environmental hazards threatened their well-being (Leonard, 1989).

To date, rural poverty has exceeded urban poverty. However, as the world's population becomes more urbanised (as seen in Chapter 5), poverty is also becoming increasingly an urban phenomenon. The World Bank estimates that, by the turn of the century, half of the developing world's absolute poor will be in urban areas (World Bank, 1990a). Cities, for historical reasons, have often been located on prime agricultural land or within valuable ecosystems near rivers, lakes or coasts. Currently, these areas remain valued sites, among other things for housing and tourism development. In the developing world, it is estimated that almost half a million hectares of arable land are being converted to urban uses annually (World Resources Institute, 1996). As a result, agricultural production often becomes more intense to

**Plate 2.2** *Lands cleared for agriculture using fire, Indonesia*

Source: Rebecca Elmhirst, University of Brighton.

compensate for losses and/or is pushed into more unsuited areas, becoming potentially more damaging in turn. Furthermore, urban developments expand onto steep slopes or wetland areas and threaten remaining natural vegetation such as mangrove forests.

Whilst the threat of urban poverty to valued ecosystems is clear, it is also evident that the poor are the victims of environmental decline. For example, an estimated 600 million urban residents of developing countries have their lives threatened every day by the health impacts associated with the inadequate provision of quality water supplies, sanitation and sewage disposal, and the lack of health and emergency services (Hardoy *et al.*, 1992a). Furthermore, there are many recent cases of poorer households suffering more losses than wealthier residents when heavy rainfall causes landslides or when earthquakes strike urban environments (Hewitt, 1997). In such circumstances, money tends to afford access to the resources on which safety depends, to the building technologies and the early warning systems, and also to relief services and insurance compensation.

However, this evidence for the close association between poverty and the environment should not distort an understanding that the linkages are complex and rarely direct in rural or urban settings. There is a need, for example, to search for the deeper underlying causes of poverty and, in turn, local-level environmental decline. In subsequent chapters, a whole host of factors, including tenure arrangements, capital assets and gender relations, for example, are seen to influence the decisions poor people make and to mediate the relationship between poverty and the environment. Regularly, particular groups (not necessarily the poorest) may be especially vulnerable to environmental and other changes, even in geographical areas beyond those which would be defined as most hazardous in terms of their physical or acquired characteristics.

In summary, sustainable development in the future will require a commitment to overcoming poverty through a focus on the welfare issues of the poorest sectors of society, particularly in the developing countries. Their environmental concerns and their development needs, in stark contrast to 'wealthy' or 'Northern' priorities, are associated with securing the most basic levels of economic and social well-being. Poverty denies millions of people basic rights in the short term and the potential to achieve development aspirations in the future. The increasing concentration of poverty in some of the world's most vulnerable locations means that only through a global commitment to addressing these interdependent concerns of the poor will remaining valued ecologies be conserved.

**Plate 2.3**  *Cleaning up the impacts of past development: recycling in the Netherlands*

Source: Gordon Walker, Staffordshire University.

## The human cost of contemporary development

The need for sustainable development in the future is also confirmed by the human cost of patterns and processes of development to date.

At present, over 12 million children under the age of five die annually in the developing world and this excludes deaths through famine (UNDP, 1996). 'Wealth determines health' is a phrase which is often used to explain the spatial pattern of ill health and premature death at various levels. A child born in the developing world today, for example, is often over ten times more likely to die before the age of one than a child born in wealthier nations (see Table 2.5). The poorer groups within a country or region also tend to suffer more premature deaths than richer groups. Figure 2.6 shows the differences in infant mortality rates in England and Wales between classes distinguished on the basis of income.

Good health, however, is not assured by higher levels of national wealth. In Table 2.5, it is seen that Sri Lanka has lower levels of infant mortality than Brazil, despite also having a lower GNP. What is not shown in the table is that Sri Lanka has a more even distribution of wealth, higher completion rates for primary schooling, higher enrolment in secondary education and greater levels of government spending on health (Wilson, 1992). Good health, therefore, is also

**Table 2.5**  *Percentage of total children who will not reach one year of age, selected countries*

| Country | GNP per capita (dollars 1995) | Infant mortality | |
|---|---|---|---|
| | | 1980 | 1995 |
| Niger | 220 | 15 | 12 |
| Ghana | 390 | 10 | 7 |
| India | 340 | 12 | 7 |
| Zimbabwe | 540 | 8 | 6 |
| Indonesia | 980 | 9 | 5 |
| Peru | 2,310 | 8 | 5 |
| Brazil | 3,640 | 7 | 4 |
| Ecuador | 1,390 | 7 | 4 |
| Sri Lanka | 700 | 3 | 2 |
| USA | 26,980 | 1 | 1 |
| UK | 18,700 | 1 | 1 |
| Japan | 39,640 | 1 | 0.5 |

Source: World Bank (1997b).

**Figure 2.6** *Infant mortality during first year of life by social class in England and Wales, 1992*

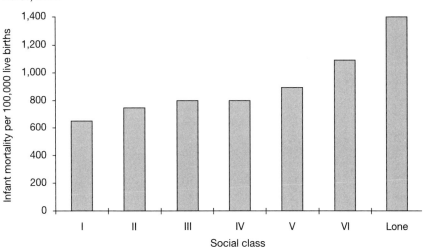

Note: 'Lone' refers to births registered in mother's name only. Social classes from I (professional) to VI (unskilled/manual).

Source: Maxwell, 1998.

about the alleviation of poverty, including through the provision of education services.

Children are amongst the poorest groups in all societies. Of the 12 million child deaths per year in the developing world, 3 million die of diarrhoeal diseases resulting from poor quality water and standards of sanitation (Satterthwaite *et al.*, 1996). A further 5 million die of diseases such as whooping cough or measles, infections which in the presence of malnutrition become major killers in the developing world but which are now relatively minor causes of ill health in the developed nations. An estimated 2 million children die annually from malaria, and many others from a variety of illnesses often associated with the debilitating presence of malnutrition and worms.

Children also suffer most from the effects of pollution (Satterthwaite *et al.*, 1996). A resting three-year-old consumes twice as much oxygen and therefore twice the pollution weight for weight as a resting adult. In addition, children's underdeveloped kidneys, livers and enzyme systems are less able to process such pollutants. Children's activities may also place them at particular risk: babies instinctively suck much of what they pick up in their hands and young children play in dangerous places such as streets or waste tips, aggravating the risk of contact with contaminated sources. Young children are responsible for very little pollution but are themselves extremely vulnerable to the pollution effects caused by others and, once again, it is impoverished children who are most at risk.

Whilst inequalities in wealth go a long way towards explaining spatial patterns of ill health at several levels, other factors are also important. For example, more children die before the age of five in West Africa than in East Africa. This has been linked to the prevalence of malaria in the former for which there is no reliable vaccine (Hill, 1991). In this instance, it is the particular environmental conditions of West Africa, rather than regional differences in wealth, which explain such patterns. Gender is another factor which cuts across inequalities in wealth to influence health. In many areas of the developing world, for example, male children are valued more highly than female, leading to a number of practices which may result in premature deaths of girls as illustrated in Box E. The UNDP has concluded that 'no country treats its women as well as men' (1993: 16). Figure 2.7 illustrates what happens to the Human Development Index (HDI) when adjusted for gender disparities (including differences in male and female life expectancy). In 1993 for example, Japan had the highest HDI of 0.983, with Niger featuring at the bottom of the HDI list with a score of 0.204. In some cases, gender discrimination is substantial, as in the Republic of Korea, where the gender-disparity-adjusted HDI is fully 35 per cent lower than the HDI (UNDP, 1993).

**Figure 2.7** *Difference between HDI and gender-disparity-adjusted HDI*

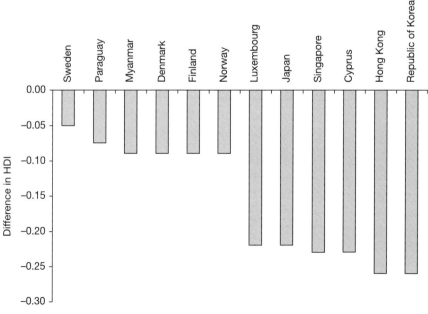

Source: UNDP, 1993.

## Box E

___

### *India's missing millions*

> India has 40 million fewer women than it would have, if the sexual balance had been left to nature.
>
> (Popham, 1997)

For centuries, various techniques have been used in India to do away with baby girls who are considered a financial burden. Although dowries are forbidden under the Anti-Dowry Act, the law is rarely enforced and even in the poorest areas of India, the bride's family is often expected to pay as much as £500 and to provide a lavish wedding feast, which can ruin poorer families or those with several daughters (Lees, 1995). In contrast, sons are considered saviours in that they carry on the family line and are indispensable economically. Although female babies are generally born stronger, statistics from hospitals reveal that girl children get less priority in nutrition and health care, and are breast-fed less than boys. Within a month of birth, the death rate of India's female babies is much higher than that of its males (Chissick, 1990). Girls are an estimated four times more likely to suffer from acute malnutrition in India and boys forty times more likely to be taken to hospital (Todaro, 1997: 158).

Although female infanticide has been banned in India for more than a century, the advent of the amniocentesis test has enabled the practice of selective abortion of female foetuses. Clinics in major cities advertise pre-natal sex-determination tests and, in 1988, a sample survey carried out in Bombay revealed that, of 8,000 abortions carried out, 7,999 were female foetuses. This prompted an Act in Maharashtra state in which Bombay lies, to regulate the use of pre-natal sex determination (the only state in India to do so), but many clinics continue to use the technique for other than the purpose of identifying genetic defects for which it was designed (Chissick, 1990).

In rural areas, women's status is particularly poor and female deaths are higher. In 1985, UNICEF funded a survey of midwives working in rural areas and found that female infanticide was extremely common:

> Dukni, 60, is a *dai*, a traditional midwife from Bihar, Northeast India. Every month, on average, she delivers six babies and kills three of them because they are girls and the families do not want them. She prefers the quickest, cleanest methods. Usually she strangles the baby with a length of rope, poisons her with fertiliser or puts a lump of black salt in her throat. Sometimes she snaps her spine by bending it backwards or suffocates her by stuffing her into a large clay pot and closing the lid. The baby is usually dead within two hours. Dukni then wraps the body in an old cloth and carries it away under her sari. She either throws it into a river, buries it or leaves it in undergrowth knowing it will quickly be found by wild animals.
>
> (Lees, 1995)

UNICEF also estimated that a further 5,000 women every year were burnt to death by their in-laws in retribution for their families' failure to provide sufficient dowry (Popham, 1997).

In 1997, the Indian government launched a £12.4 million scheme to reward couples for having daughters. Families who earn less than £190 per year who produce a daughter will be rewarded with a payment of £9 and also receive financial incentives to encourage them to send their daughters to school. They also announced a ban on tests to determine whether an expectant mother is carrying a boy or a girl child. The scheme has been given a cautious welcome by those who work with girls and women in poor communities in India; it is the first time the authorities have pledged money, but much more is needed in terms of ensuring a proper educational infrastructure in the rural areas for example, and further incentives to advance the position of women in Hindu society.

Table 2.6 *Expenditure on health and education in the 1980s in selected countries in Africa and Latin America (health and education spending as percentage of GNP)*

| Country | Spending on health and education | | Percentage change |
|---|---|---|---|
| | 1981 | 1990 | |
| Kenya | 8.10 | 7.91 | −0.91 |
| Tanzania | 5.90 | 2.93 | −2.97 |
| Nigeria | 1.6 | 1.00 | −0.60 |
| Botswana | 11.04 | 10.55 | −0.49 |
| Sierra Leone | 5.80 | 1.55 | −4.25 |
| Bolivia | 4.01 | 3.82 | −0.19 |
| Chile | 6.45 | 5.25 | −1.20 |
| Mexico | 4.18 | 2.91 | −1.27 |
| Peru | 3.35 | 2.13 | −1.22 |
| Ecuador | 6.50 | 4.56 | −1.94 |

Source: Stewart (1995).

The extent of ill health and premature death, particularly amongst children, is evidence that current development patterns and processes are not meeting the needs of current or future generations. Spatial inequalities in health have been seen to exist at various scales and show strong links with the distribution of wealth internationally between the developed and developing worlds, but also within nations. However, in much of the developing world, with few exceptions, the recent trend has been for declining overall expenditures on health and on education, as seen in Table 2.6, despite the mounting challenges including that of HIV infection. Furthermore, in the early 1990s, the introduction of structural adjustment programmes forced cut-backs in government spending, which often hit these service areas hardest. Such trends clearly threaten to produce an increasingly ill-educated as well as unhealthy young population.

In addition, health has been seen to be closely tied to the physical environment, particularly in the developing world where poverty is so widespread and entrenched. Elimination of poverty and the provision of basic welfare needs for large numbers of people in the developing world must therefore form an essential basis for future development patterns and processes if this human cost of unsustainable development is to be avoided.

Plate 2.4 *AIDS awareness campaign in Zambia*

Source: David Nash, University of Brighton.

## The environment cannot cope

Perhaps the starkest realisation of the need to find new patterns and processes of development has come from an improved understanding of the environmental unsustainability of contemporary development. At the time of the Stockholm conference, the primary environmental problems tended to be national: 'the environmental sins of one nation did not generally impinge upon other nations, let alone upon the community of nations' (Myers and Myers, 1982: 195). Ten years later, further environmental problems which affected many nations, such as acid rain or nuclear waste, were recognised. More recently, ash carried by local atmospheric patterns from the forest fires burning in Indonesia led to severe problems of air quality and impacts on health and production in Malaysia and Singapore, for example. However, in the past twenty years, there has been increasing recognition of a series of environmental problems which now affect the global community as a whole. Included in this 'supranational' category of contemporary environmental issues are the destruction of ozone in the stratosphere and the problem of 'global warming': the heating up of the Earth owing to the accumulation of 'greenhouse gases' (particularly carbon dioxide) in the upper atmosphere. Such issues 'deserve collective measures on the part of humankind as a whole' (Myers and Myers, 1982: 200).

The temperature of the Earth is regulated by 'greenhouse gases' in the atmosphere which control the re-radiation of solar radiation back into space. These gases serve to keep the Earth some 33°C warmer than it would otherwise be. However, the concentration of greenhouse gases has increased during the past century and the average temperature of the Earth's surface is considered to have risen by at least 0.5°C over the same period, as shown in Figure 2.8. Currently, carbon dioxide accounts for over half of the warming or 'forcing effect' in the atmosphere (Barrow, 1995).

The major source of carbon dioxide to date has been the burning of fossil fuels (Foley, 1991). Figure 2.9 illustrates the varied sources of carbon dioxide (both natural and human induced) and the routes for its transportation. Human activities are now placing more carbon dioxide into the atmosphere more quickly than the natural sinks of the gas, the oceans and green vegetation, can remove through processes of diffusion and photosynthesis. Processes of deforestation have been an important factor in rising concentrations of atmospheric

**Figure 2.8** *Global surface temperature rise, 1880–1990*

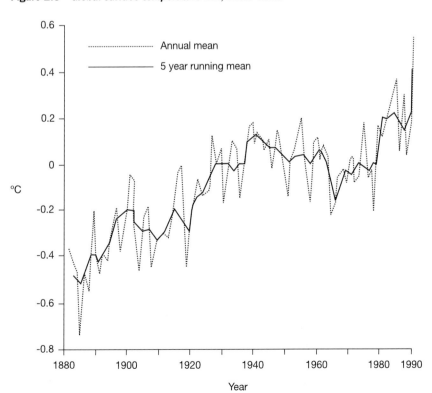

Source: Rees (1998).

**Figure 2.9** *The carbon cycle*

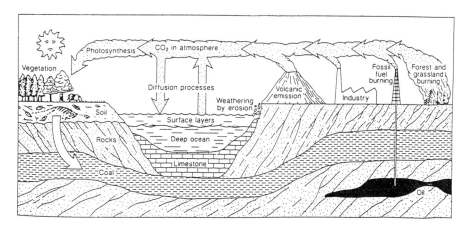

Source: Pickering and Owen (1994).

carbon dioxide; the removal of trees itself reduces the uptake of carbon dioxide and the associated burning of logs and biomass sources contributes further to the production of carbon dioxide (Potter *et al.*, 1999).

In future, the contribution of carbon dioxide to forcing global warming is likely to decline in relation to the importance of further greenhouse gases, namely chlorofluorocarbons (CFCs), the synthetic chemicals used in various processes and products including within refrigeration and as propellants. The key feature of CFCs is that they remain active in the atmosphere for much longer periods than carbon dioxide, before being destroyed in the stratosphere by ultraviolet radiation. Therefore, although the production of CFCs has been declining in the 1990s (Brown, 1996), atmospheric concentrations will continue to increase for centuries.

It is widely estimated that the global mean temperature will rise by 3°C during the early decades of the next century (Reading *et al.*, 1995: 353). The major effects of such warming will relate to water resources: through the rise of sea levels as a result of the thermal expansion of the oceans, the melting of glaciers and ice-sheets, and increased precipitation via enhanced evaporation from warmer seas (Potter *et al.*, 1999). Indeed, there is a veritable industry of activity currently taking place which attempts to link environmental events such as the El Niño phenomenon to global warming. What is certain is that the impact of climate warming will not be distributed evenly over the globe. Figure 2.10, for example, summarises the regional impacts

**Figure 2.10**  *IPPC estimates of regional climate changes by 2030*

....................................................................................................................

**Central North America**

The warming varies from 2 to 4 °C in winter and 2–3 °C in summer. Precipitation increases range from 0 to 15 per cent in winter whereas there are decreases of 5–10 per cent in summer. Soil moisture decreases in summer by 15–20 per cent.

**Southern Asia**

The warming varies from 1 to 2 °C throughout the year. Precipitation changes little in winter and generally increases throughout the region by 5–15 per cent in summer. Summer soil moisture increases by 5–10 per cent.

**Sahel**

The warming ranges from 1 to 3 °C. Area mean precipitation increases and area mean soil moisture decreases marginally in summer. However, throughout the region, there are areas of increase and decrease in both parameters throughout the region.

**Southern Europe**

The warming is about 2 °C in winter and varies from 2 to 3 °C in summer. There is some indication of increased precipitation in winter, but summer precipitation decreases by 5–15 per cent, and summer soil moisture by 15–20 per cent.

**Australia**

The warming ranges from 1 to 2 °C in summer and is about 2 °C in winter. Summer precipitation increases by around 10 per cent, but the models do not produce consistent estimates of the changes in soil moisture. The area averages hide large variations at the sub-continental level.

....................................................................................................................

Source: Houghton *et al.* (1997).

estimated by the Intergovernmental Panel on Climate Change. It is seen that the predicted warming and impact on soil moisture vary between and within continents.

Through the disruption of sea levels, ocean currents and the constituent gases of the atmosphere, climate warming is the archetypal global environmental issue: 'a molecule of greenhouse gases emitted anywhere becomes everyone's business' (Clayton, 1995: 110). As such, its status arguably at the top of the international environmental agenda, as will be seen in Chapter 3, is perhaps easy to understand. However, it is also evident from the preceding sections that the world's environmental challenges extend far beyond these issues of the 'global commons'.

## Global challenges for the future

Experience of suffering from pollution and environmental degradation is 'nowadays almost universal' (Yearley, 1995: 146). Much of present-day pollution has an international impact and local pollution problems are repeated the world over. However, the brief detail given in the previous sections as to the nature of environmental concerns between and within nations has given some sense of the highly unequal and uneven experience and impact of environmental decline. As such, 'it would be wrong, therefore, to assume that responses to pollution problems are globally harmonious' (ibid.: 147). The following sections consider a number of important challenges for the global community in terms of managing environmental change into the future.

## Questions of responsibility and response

The current interdependence of peoples and environments throughout the world is seen starkly in the case of supranational environmental problems. The need for a co-ordinated international response to such environmental issues is self-evident. However, in the moves to determine the response to such problems, the question of responsibility is inevitably raised. If change is to be implemented, particularly where such change involves (as it usually does) expense and/or compromise of some sort, there is substantial debate as to whether these costs should be borne equally by all or whether there should be some adjustment according to relative responsibility, if this can be ascertained (and also with respect to ability to pay).

One of the most fundamental challenges for ensuring sustainable productive activities in the future stems from the fact that pollution effects are often spatially and temporally removed from the site of production, as already noted. As such, it is often extremely difficult to ascertain the responsibility for pollution. Box F considers aspects of the debate which continues in respect to the responsibility for global warming. The 'battle of statistics' seen in the debate is an extremely important one, since potential solutions to the problem will be argued on different figures.

One of the favoured mechanisms for minimising pollution generally is the 'Polluter Pays Principle' (PPP), defined in Figure 2.12. The PPP is seen to rest on the assumption that it is possible to identify who is responsible for the production of pollution. In addition, it depends on

## Box F

---

### Responsibility for global warming under debate

Based on scientific understanding of the role of carbon dioxide in particular in forcing global warming and the important role of the burning of fossil fuels in these processes, there have been many attempts to establish relative responsibility internationally for environmental damage. In the main, it has been identified that carbon dioxide emissions from the industrialised nations have, to date, far outweighed the contribution from the developing world, as shown in Table 2.7. In terms of all greenhouse gases, Kelly and Granwich (1995) assert that the industrialised nations (with approximately one-quarter of the world's population) have been directly responsible for over half of the emissions causing global warming (p. 92).

Table 2.7  *'Top fifteen' carbon culprits*

| Country | Total annual emissions (million metric tonnes) | Per capita |
|---|---|---|
| United States | 4,881 | 19.1 |
| China | 2,668 | 2.3 |
| Russian Federation | 2,103 | 14.1 |
| Japan | 1,093 | 8.8 |
| Germany | 878 | 10.9 |
| India | 769 | 0.9 |
| Ukraine | 611 | 11.7 |
| United Kingdom | 566 | 9.8 |
| Canada | 410 | 14.4 |
| Italy | 408 | 7.2 |
| France | 362 | 6.3 |
| Poland | 342 | 13.5 |
| Mexico | 333 | 8.9 |
| Kazakhstan | 298 | 17.6 |
| South Africa | 290 | 7.5 |

Source: World Bank (1997b).

However, in 1990, the World Resources Institute (the influential Washington-based research group whose environmental data are used regularly by policy-makers and other researchers) published quite different data concerning the contribution of individual nations to global warming, based on carbon emissions from sources beyond solely fossil fuel combustion. The conclusion was that responsibility was evenly shared between industrialised and developing regions, as shown in Figure 2.11a.

**Figure 2.11** *Responsibility for net emissions of greenhouse gases as calculated by (a) the World Resources Institute and (b) the Centre for Science and Environment*

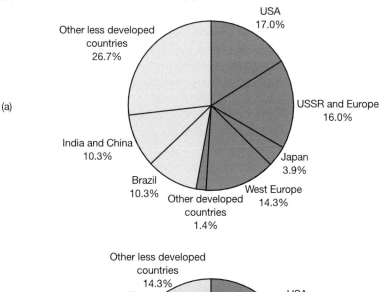

(a)

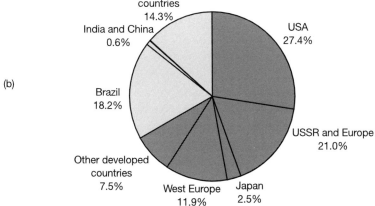

(b)

Source: Barrow (1995).

In turn, these findings have been much criticised. For example, McCulley (1991) notes that the European Community is defined and counted as one 'industrialised country'. Furthermore, in calculating a single 'greenhouse index' based on one year's emissions of greenhouse gases into the atmosphere, the WRI is able to exaggerate the contribution of the developing countries, through choosing not to consider responsibility in terms of historical contributions to emissions. No account is taken of the distribution also of the world's 'sinks' for greenhouse gases. The uncertain nature of much data (such as those concerning carbon dioxide releases from land use changes) is not made explicit, nor is there any attempt within the calculations of the World Resources Institute to differentiate between emissions for 'luxury' (such as vehicle use) and 'survival' purposes (as in the case of methane from paddy fields).

Figure 2.11b illustrates the divergence in terms of responsibility calculated by the WRI and an alternative suggested by the Centre for Science and Environment, based in India. The latter has argued that comparisons should be made between countries' emissions on a per capita basis, since 'every person has a moral right to air' (Barrow, 1995: 90). However, such a calculation would favour countries with large, predominantly young populations over those with slow-growing populations (largely the 'developed' countries).

---

**Figure 2.12** *The Polluter Pays Principle (PPP)*

The principle originates from the proceedings of the UN Conference on the Human Environment in Stockholm, 1972. The principle is that the cost of preventing pollution or minimising environmental damage due to pollution should be borne by those responsible for the pollution. Measures such as taxing processes which generate pollution (for example, the use of leaded fuels) or payments for licences to emit certain levels of pollutant (such as in waste management in the European Union) are in line with the principle that the polluter should pay.

---

being able to 'cost' the damage. The challenge of costing the impact of pollution is seen clearly in the case of 'acid rain', which refers to the abnormally low pH of some rainfall resultant from the concentration of primarily sulphur dioxide as a result of burning fossil fuels. Whilst it may be possible to cost the damage to trees in terms of the loss of timber resources, for example, it is much more difficult to assign a market value to the loss of diversity of flora and fauna supported by such forests, or to their decreased amenity value with regard to recreational opportunities lost for the local population as a result of acid rain.

Even where it is clear who is responsible for pollution, as in the case of the breakdown of the nuclear reactors at Chernobyl in Ukraine in 1986, the impacts can spread far beyond national boundaries, as shown in Figure 2.13. Furthermore, the impacts of Chernobyl were experienced unevenly over space and time: 'the soil character and farming practices of Cumbria and North Wales meant that livestock have been contaminated for years; elsewhere in the United Kingdom the most obvious impacts passed in weeks' (Yearley, 1995: 156). Moreover, there is often substantial disagreement within the scientific community over the nature of the link between the pollutant and the ill health of ecosystems and humans; there is much that is not known, for example, regarding the thresholds above which nuclear wastes are hazardous to humans and concerning the length of time taken for such products to become harmless. It is this problem of a lack of

**Figure 2.13**  *The progress of the Chernobyl plume, 1986.*

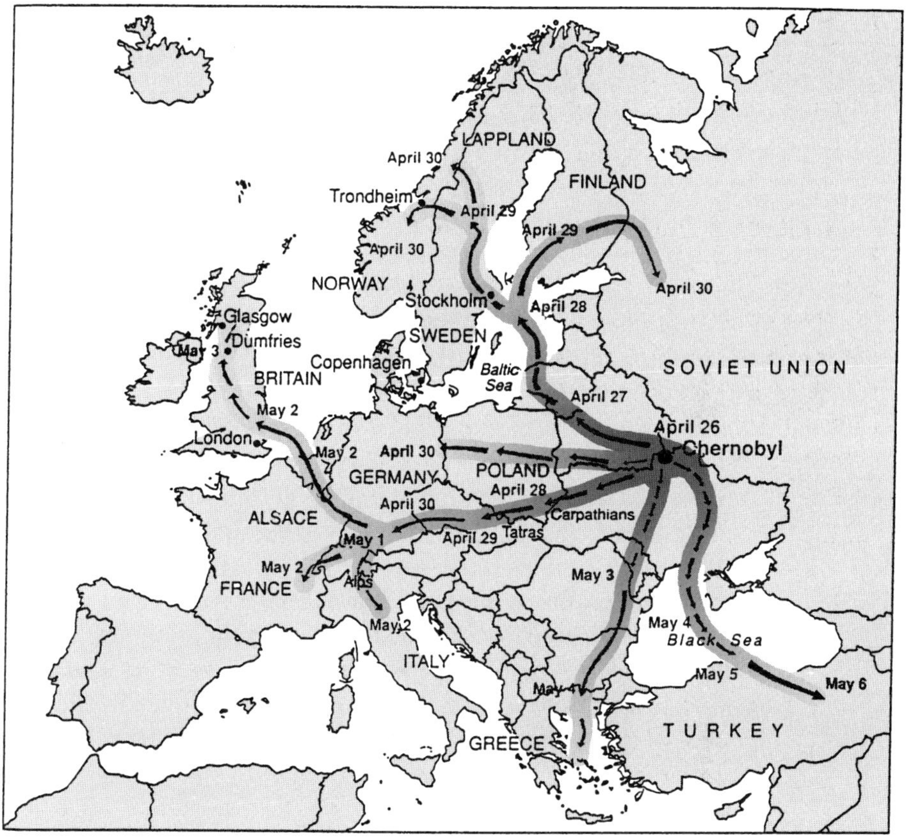

Source: Allen and Hamnett (1995b).

understanding regarding the links between cause and effect in the
production and impact of pollution which represents a major
challenge for ensuring that future productive activities at all levels are
sustainable.

## The power to respond

It has already been noted that the environmental concerns of the poor
are related to survival itself. This applies as much to poor countries as
it does to their low-income populations. Clearly, whilst it is unrealistic
to expect poor people to conserve resources for the future when they
are struggling for survival, the governments of these countries, in
turn, have very scarce economic resources for any activities outside the
provision of basic human needs. Rees (1997) suggests that the real

environmental challenge for the global community in the future will be 'to achieve consensus over what constitutes an equitable allocation of environmental protection costs between social groups, sectors within national economies and, above all, between nations' (p. 461).

Even global problems such as climate change have uneven impacts, as seen in Figure 2.10. In part, these varied impacts are a function of geographic location, particularly in relation to major atmospheric and oceanic circulations, for example. Whilst low-lying nations such as the Netherlands and Bangladesh are vulnerable to the impacts of sea-level rise, the consequences for human society in each country are very different due to factors including wealth. Generally, in wealthier countries, pollution impacts can be managed more easily, whether this comes through the provision of safe sewerage, systems of early warning and disaster response, or through the development, enforcement and monitoring of legislation. The varied pattern of environmental planning regulations in the late 1970s between countries such as the US and nations of the developing world, for example, was an important factor in encouraging multi-national firms to seek 'pollution havens' where laws concerning location and the control of emissions or wastes were less demanding. Fundamental differences in power in the capitalist economy therefore enabled people in one part of the globe to displace their environmental problems onto other parts and save money and trouble in so doing (Yearley, 1995).

The Brundtland Commission estimated that, in the early 1980s, the cost for developing countries of bringing environmental legislation up to US standards was in excess of US$5 billion. 'The sum today would be considerably larger' (LeQuesne and Clarke, 1997: 171). The international trade in hazardous waste considered in Box G is a clear illustration of where the lack of economic power on the part of developing nations constrains sustainable development. Although there are now international conventions aimed at regulating trade in hazardous wastes (most importantly the Basel Convention which came into force in 1992), the trade continues. Fundamentally, receiving countries are so in need of income that they receive the pollutant 'without even the benefits of hosting the industrial processes which cause it' (Yearley, 1995: 165). Their lack of economic power also extends to their inability to monitor illegal trade into their country or to enforce penalties on individuals and companies that conspire to extend such trade. In addition, there continues to be exemptions to the convention, for example, concerning 'recovery operations'. Until 1998, a wide range of so-called 'recycling' exports

## Box G

---

### *The export of hazardous waste*

Already jeopardised by global environmental changes, erosion, famine, and deforestation, the Third World is currently being invaded by extensive exports and dumping of tons of hazardous waste from industrialised countries . . . For the millions of people in Third World countries, the sometimes legal, sometimes illegal importation of hazardous waste represents an unprecedented threat to their health and their environment.

(Sanchez: 1994: 137)

Hazardous waste is defined as waste which, if deposited into landfills, air or water in untreated form, will be detrimental to human health or the environment. It includes toxic, flammable, explosive and nuclear materials. Reliable statistics on the international movement of hazardous materials are hard to come by, because of the illegal nature of much trade and the varied standards or definitions of waste used in different countries. However, 175 million tonnes of hazardous waste were offered on formal world markets between 1986 and 1991 (*Guardian*, 1992). Ten million tonnes of such wastes

**Figure 2.14   *The West African trade in hazardous waste***

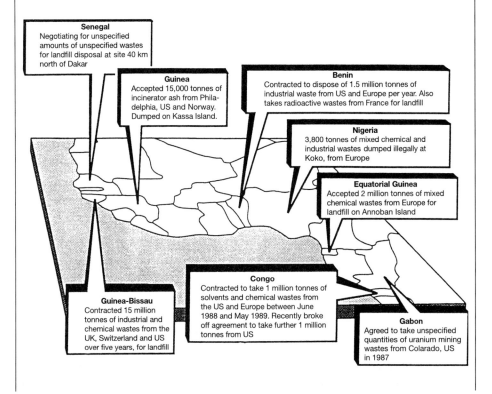

**Senegal**
Negotiating for unspecified amounts of unspecified wastes for landfill disposal at site 40 km north of Dakar

**Guinea**
Accepted 15,000 tonnes of incinerator ash from Phila-delphia, US and Norway. Dumped on Kassa Island.

**Benin**
Contracted to dispose of 1.5 million tonnes of industrial waste from US and Europe per year. Also takes radioactive wastes from France for landfill

**Nigeria**
3,800 tonnes of mixed chemical and industrial wastes dumped illegally at Koko, from Europe

**Equatorial Guinea**
Accepted 2 million tonnes of mixed chemical wastes from Europe for landfill on Annoban Island

**Guinea-Bissau**
Contracted 15 million tonnes of industrial and chemical wastes from the UK, Switzerland and US over five years, for landfill

**Congo**
Contracted to take 1 million tonnes of solvents and chemical wastes from the US and Europe between June 1988 and May 1989. Recently broke off agreement to take further 1 million tonnes from US

**Gabon**
Agreed to take unspecified quantities of uranium mining wastes from Colarado, US in 1987

were exported during the same period, largely from the industrialised countries to the developing world. The main receiving areas at that time were the Caribbean and Central and South America, with huge growth in shipments to the former Soviet states. Figure 2.14 highlights the nature and location of recent movements of hazardous wastes in West Africa.

Yearley (1995) refers to hazardous waste trading as 'a business which had seemed set to vanish' (p. 167) in the 1990s. In 1988, for example, the Organisation of African Unity passed a resolution calling for a ban on the importation of hazardous wastes to the continent. Individual countries had also reversed earlier agreements to import wastes. Under the 1992 Basel Convention convened by UNEP, it is recommended that the importing country should show evidence of its capacity to deal with the particular waste prior to any trade. However, such regional and international conventions are proving inadequate controls when set against the economic incentives for developing countries to continue to receive hazardous wastes.

The initial explosion in the 1980s in the export of hazardous waste and its continuation can be considered to be a product of two main factors: the uneven progress in the development of environmental standards throughout the world and the rise of neo-liberal economic thinking on the part of government leaders in more developed nations, and particularly within international institutions such as the World Bank. In broad terms, consumer demand for clean industry has been an important factor in the development of more and more stringent rules concerning the treatment, storage and disposal of toxic waste materials in industrialised societies. The added cost associated with adherence to these regulations, as well as the lack of land in countries such as Britain to receive the growing volume of wastes, has led to companies taking the easy option: the dumping of toxic wastes in poor countries with large areas of land and few restricting regulations on disposal.

Raising export earnings is central to the structural adjustment programmes promoted by the World Bank and International Monetary Fund (as seen in Chapter 1), such that debtor countries are under pressure to attract investment 'even if that meant accepting dirty industry or building incinerators to deal with wastes of the industrialised North' (Yearley, 1995: 167). In addition, the cuts in government spending regularly required within such policy reforms serve to reduce the capacity for pollution control measures to be enforced and monitored further. In 1992, the vice-president of the World Bank, Lawrence Summers, was quoted as supporting trade in hazardous materials, citing Africa as a particular example of an 'under-polluted' region and referring to trade in air pollution and waste as 'world-welfare enhancing' (*Guardian*, 1992). The suggestion was that, like any other commodity, hazardous waste should be allowed to move across borders within a free market economy.

The attractions to individual companies and governments of poor countries alike of importing toxic wastes in terms of short-term monetary gain are clear. For example, two British firms offered US$120 million per year to Guinea-Bissau to bury industrial waste material, which was equivalent to the annual GNP of that country (Third World Network, 1989). In the Congo, a number of individuals agreed to import one million tonnes of industrial waste from Holland which would have made US$4 million over three years (ibid.). The amounts paid to the receiver of hazardous materials are negligible, however, compared with the savings made on storage and disposal in the country of origin. The sum of US$40 per tonne was paid to Guinea importers of toxic industrial ash from Philadelphia. Storage in the USA in compliance with government regulations would have cost the company US$1,000 per tonne.

Awareness is growing of the intractable nature of the toxic waste problem; the difficulty of making toxic wastes safe and the time span over which many materials, particularly nuclear wastes, remain dangerous are now more fully appreciated. The disposal of such products presents a substantial technological as well as political challenge. The economic, health and environmental costs of past mistakes are only just being realised. In the USA in 1980, the President's Council on Environmental Quality estimated that the cost of cleaning up existing dump sites in the country would be US$28–55 billion (Third World Network, 1989). In the UK, the cost of dismantling and disposing of redundant nuclear installations is estimated to be £20 billion. In the case of the latter, a site for the necessary deep repository is still to be found; the political considerations of where this should be are inevitably as difficult as the scientific and geological factors (ibid.). The developing nations are even more ill-equipped than the industrialised countries to deal with toxic wastes. Very rarely is the awareness or technology available in these countries for handling the long- or short-term dangers of these materials.

were able to persist, many of which were highly polluting, left toxic residues or were dangerous to workers (Puckett, 1994).

The impact of environmental degradation on society, between sectors within national economies and the costs for particular social groups, will vary also according to differences in social organisation. Under a socialist modes of societal and economic organisation, for example, control over pollution strategies tends to be highly centralised, giving little opportunity, in turn, for the monitoring or control of pollution by local authorities or citizens' groups (see Johnston, 1996, for a discussion of these issues). This is in contrast to the situation under a capitalist mode in which such responsibility is often decentralised to regional and local authorities. This is the case in the UK with many aspects of pollution control, for example, where decision-making is based on consultative procedures between ministers, various government advisory bodies and selected public advisory groups and individual consultants. Increasingly, representatives of environmental groups are being incorporated into the decision-making process too. This occurs through their co-option on to advisory boards and through public lobbying on environmental issues. In contrast, in the former USSR (prior to its break-up into independent states), most national policy-making of any importance took place in the Politburo and there was little recourse to the opinions of state agencies or local environmental needs. There was also no political mechanism to involve public groups or advisers within the decision-making process. The concentration of public environmental activities in a small number of government-sponsored groups also effectively minimised the prospects for any public pressure for environmental conservation. For the vast majority of the public, their awareness of environmental destruction, even in their local area, was restricted by the secrecy of information in the USSR 'pre-*glasnost*'.

It is now recognised that no one system of production or administration of environmental control has the monopoly on conservation. What is realised, and will be illustrated more fully in the following chapters, is that progress towards sustainable development very regularly depends on enabling individuals and local community groups to become more powerful in their control over resources and the environment. It also depends on much greater financial and technical assistance to the developing world to assist it in environmental protection. 'At present the greatest single barrier inhibiting the implementation of controls capable of tackling environmental change is the SEMPY (somebody else must pay) syndrome' (Rees, 1997: 461).

## Questions of sovereignty

A recurring issue and an important challenge for environmental management in the future concerns the questions of a country's sovereignty and the right to decide how its resources (including that of its people and their labours) should be used. Clearly, such issues are also closely related to those of the location and level of power as discussed previously.

In Chapter 1, it was noted that the initial hesitancy of developing nations to participate in the environmental debates of the 1960s was due in part to concern over the loss of control over their own development. With the rise of global governance throughout the 1990s, many of these same fears, on the part of the more developed as well as developing nations, resurfaced within international negotiations to determine actions for sustainable development. As will be seen in Chapter 3, many actions towards sustainable development undertaken in recent decades by the international community have involved the setting of standards for nations to follow. The idea that national standards should be a matter for international debate is fundamentally in contradiction to the notion of a country's sovereign right to look after the environment in a manner which they assess to be best (Cairncross, 1995). There were fierce debates, for example, at the 1997 UN climate meeting in Kyoto, Japan, as each participating nation negotiated its 'quota of pollution'. The extremely lengthy proceedings of meetings of the newly created World Trade Organisation are also seen to be, in part, a function of the concern of member states to apply their own rules, including those of environmental standards in production.

Fundamentally, international institutions themselves are created by states as a means of achieving 'collective objectives that could not be accomplished by acting individually' (Werksman, 1996: xii). By definition, therefore, the conservation of the global environment demands some devolution of sovereign power, and the success or otherwise of international institutions working towards this goal depends on the willingness of those states to make such investments. However, critical evaluation is still needed regarding the nature of those goals. There remain fears in the developing world in particular that global environmental objectives are being set according to the agenda of countries in the North. As such, priority is still given to issues of global climate change, deforestation and species extinction, for example, which are quite different to the environmental problems of the South, which concern, most regularly, basic standards of living and life itself.

## Conclusion

The inequitable nature of past development processes and patterns has been seen to be the main underlying reason why they cannot be sustained into the future, morally, economically and environmentally. Not only have the benefits and costs of progress to date fallen unevenly between nations and within sectors of society, but the persistence and entrenchment of poverty mean that increasing numbers of people are denied access to the resources on which future development depends and in the process are themselves a factor in the further degradation of those essential environments. Whilst inequalities in wealth have been seen to be important in understanding the challenges of sustainable development, from delivering basic health services through to financing pollution prevention and abatement technologies, it is also clear that these challenges are not solely economic, as illustrated in the case of premature deaths of girl children in India, for example.

There are evidently substantial new challenges for international relations. Whilst some degree of collective action has been fundamental to the survival of human societies throughout the history of civilisation, people and places today are connected in many more diverse and far-reaching ways (although at various intensities) and humanity now depends on co-operative interactions which go far beyond those of original societies in terms of both number and complexity. In the immediate term, for example, the challenges of sustainable development have been seen to include many different nation states in negotiation over the sharing of water use and in the location of potentially environmentally degrading production or products. States are also negotiating within newly created international fora in an attempt to deal with the challenges of sustainable development over the longer term, such as with respect to CFC reduction targets. However, the political challenges of sustainable development extend to individuals: the choices we make concerning our own consumption levels and preferences and in our own capacity to monitor and prompt change within the institutions that represent us. In summary, sustainable development is a challenge for people and planners across the globe rather than for particular institutions of development or for certain regions of the world in isolation.

## Discussion questions

* Rehearse the major arguments for poverty alleviation being the central focus for sustainable development.
* In what ways do issues of biodiversity challenge the global community?

* Select a contemporary international news item as covered in the print media. Critically consider the explanations of the cause (responsibility) and the solution (response) presented.

## Further reading

Foley, G. (1991) *Global Warming: Who Is Taking the Heat?*, Panos Publications, London.

Kirkby, J., O'Keefe, P. and Timberlake, L. (eds) (1995) *The Earthscan Reader in Sustainable Development*, Earthscan, London.

O'Riordan, T. (ed.) (1995) *Environmental Science for Environmental Management*, Longman, London.

Satterthwaite, D., Hart, R., Levy, C., Mitlin, D., Ross, D., Smit, J. and Stephens, C. (1996) *The Environment for Children: Understanding and Acting on the Environmental Hazards that Threaten Children and Their Parents*, Earthscan, London.

World Bank (1990a) *World Development Report*, Oxford University Press, Oxford.

# Action towards sustainable development

- The rise of global governance
- The nature and extent of international linkages
- Environmental pressures of debt servicing and trade liberalisation
- Reconciling the state's role in development with its role as environmental protector
- Democracy and empowerment

## Introduction

Accepting the necessity and desirability of sustainable development in the future was the essential prerequisite for the global community to start taking action. As the Prime Minister of Canada has been quoted as saying, 'statements of intent are the necessary precursors of action' (in Starke, 1990: 49). Measures taken by actors at various levels with the explicit aim of moving towards sustainable development have escalated over the past ten years. In so doing, however, the real challenges of sustainable development, those of reconciling the ambitions of various interest groups, of identifying basic versus extravagant needs and of balancing present and future development aspirations, have all become clearer. Inevitably, the practice of sustainable development is proving more difficult than professing an intention, yet there are signs of progress.

This chapter identifies some of the progress that has been made. The focus is on actions taken at various levels, from international institutions through to community organisations. In subsequent chapters, the analysis will focus on actions taken within particular sectors, urban and rural.

## International action

Perhaps our most urgent task today is to persuade nations of the need to return to multilateralism . . . after a decade and a half of a standstill or even deterioration in global co-operation the time has come for higher expectations, for common goals pursued together, for an increased political will to address our common future.

(WCED, 1987: x)

One of the primary means by which countries can confirm their co-operation within international efforts to support global environmental goals has been through their signatures to various protocols, treaties and conventions that bind international behaviour. Over a hundred such environmental treaties have been adopted since the 1972 UN Conference on the Human Environment (World Resources Institute, 1994). For example, in 1987, the United Nations Environment Programme (UNEP) brought government representatives together in Montreal to consider a protocol on substances that deplete the ozone layer. Governments representing two-thirds of global CFC use agreed to targets for the phasing out of such substances, and the 'Montreal Protocol' became effective in 1989. In 1992, signatory nations met again in Copenhagen to review the status of phase-out periods and committed themselves to an acceleration of reductions in the light of further scientific evidence regarding ozone loss. In 1995, the 150 parties to the protocol revised it once again to include controls on hydrochlorofluorocarbons which are being used as replacements for CFCs, but are also damaging ozone levels (Starke, 1997).

Similarly, in 1987, UNEP established an Intergovernmental Panel on Climate Change (IPCC) to co-ordinate research and to identify response strategies to global warming. The IPCC was instrumental in designing the UN Framework Convention on Climate Change signed by more than 150 states at the Earth Summit in Rio. Parties are obliged to aim to reduce their emissions of carbon dioxide by the year 2000 to levels lower than those in 1990. However, the Convention recognised 'common but differentiated responsibility' in terms of the relative contribution of different world regions to the problem. Many of the developing countries, for example, were not required to take on obligations beyond broad reporting. The Conference of Parties has now met three times since Rio and at the most recent meeting in Kyoto in Japan in late 1997, a legally binding agreement on emission reductions was reached.

In addition, issues of the international commons, those physical and biological systems which lie outside the jurisdiction of any particular country, but whose services are valued by society as a whole (O'Riordan, 1995: 348), have received international attention. The International Tropical Timber Organisation (ITTO) was formed in

1983, and by the end of the decade had 45 member countries who accounted for 80 per cent of these forests and 95 per cent of tropical timber exports (Starke, 1990). All members are signatories to an agreement committed to the sustainable use and conservation of tropical forests. In 1990, the ITTO announced 'Target 2000' whereby all trade in tropical timber by that date should be supplied from sustainable logging. However, many professional foresters, funding agencies and researchers remain sceptical concerning the possibility of logging (which inevitably simplifies ecosystems) being sustainable in terms of maintaining all the biological, ecological and social functions of tropical forests (Colchester, 1990).

The Convention on International Trade in Endangered Species of Wild Flora and Fauna (CITES) was established in 1973 and continues to be the major international monitor of species loss. In 1992, it added the African elephant to its list of endangered species and introduced a controversial ban on the sale of ivory in an attempt to limit poaching of these animals. At Rio, 155 states and the European Union signed the Convention on Biological Diversity. It goes further than previous conventions, such as CITES, in establishing a wider context for all biodiversity protection and for the sustainable use of the components of biodiversity (Munson, 1995). The Convention has been variously described as the most significant outcome of UNCED (Barrow, 1995) and the minimum on which the international community would agree (Tolba, cited in Munson, 1995). In particular, the USA did not sign on the suggestion of the threat that it posed to the US biotechnology industry.

At the Earth Summit, all participating countries agreed to the Rio Declaration on Environment and Development, a set of twenty-seven principles for the future conduct of nations and peoples with respect to environment and development as summarised in Figure 3.1. Agenda 21, containing forty chapters, was the 'action plan' to emerge from UNCED. It is not a legal agreement in the sense that governments are not required to follow each recommendation 'line by line'. However, it has been referred to as a 'collection of agreed and negotiated wisdoms as to the nature of the problems and relevant principles of the desirable and feasible paths . . . against which government and other actions can and will be compared' (Koch and Grubb, 1997: 455).

Also at Rio, the international community agreed to the formation of a UN Commission on Sustainable Development (CSD) as an overarching international environmental organisation. Its principal

**Figure 3.1** *The Rio Declaration on Environment and Development*

...........................................................................................................................

1  Human beings are at the centre of concerns for sustainable development. They are entitled to a healthy and productive life in harmony with nature.

2  States have the sovereign right to exploit their resources and the responsibility to ensure that such exploitation does not cause damage to the environments of other states.

3  The right to development should equitably meet the needs of present and future generations.

4  In order to achieve sustainable development, environmental protection shall be an integral part of the development process.

5  All states shall co-operate in eradicating poverty.

6  Developing countries, and especially the least developed and most environmentally vulnerable, will be given special priority.

7  States have common but differentiated responsibilities. The developed countries acknowledge their responsibility in view of the pressure they place on the global environment.

8  States should reduce and eliminate unsustainable patterns of production and consumption and promote appropriate demographic policies.

9  States should co-operate to strengthen capacity for sustainable development through exchanges of scientific and technological knowledge.

10  States shall facilitate and encourage public awareness by making environmental information widely available.

11  States shall enact effective environmental legislation.

12  States shall co-operate to promote a supportive and open international economic system.

13  States shall develop national law on liability and compensation for victims of pollution and other environmental damage.

14  States should co-operate to discourage or prevent the relocation and transfer to other states of any activities or substances that cause severe environmental degradation or harm to human health.

15  The precautionary principle shall be applied by states.

16  National authorities should promote the internalisation of environmental costs and the use of economic instruments. The polluter should, in principle, bear the costs of pollution.

17  Environmental impact assessment shall be undertaken for proposed activities that are likely to have significant adverse impact on the environment.

18  States shall notify other states of natural disasters or other emergencies likely to produce sudden harmful environmental effects.

19  States shall notify potentially affected states of activities that may have a significant transboundary environmental effect.

20  Women have a vital role in environmental management and their participation is therefore essential to achieve sustainable development.

21  The creativity, ideals and courage of youth should be mobilised to forge a global partnership in order to achieve sustainable development.

22  Indigenous people and other local communities have a vital role in environmental management and development.

23 The environment and natural resources of people under oppression, domination and occupation shall be protected.

24 Warfare is inherently destructive of sustainable development. States shall therefore respect international law providing protection for the environment in times of armed conflict.

25 Peace, development and environmental protection are interdependent and indivisible.

26 States shall resolve all their environmental disputes peacefully and by appropriate means in accordance with the Charter of the United Nations.

27 States and people shall co-operate in the fulfilment of the principles embodied in this Declaration and in further development of international law in the field of sustainable development.

function is to monitor the implementation of Agenda 21 through review of all reports from relevant organisations and programmes within the UN system. The UN itself has six main organs including the Economic and Social Council and the Security Council, plus various programmes such as the UNDP, the UNEP and the UN Children's Fund (UNICEF). A number of specialised agencies including the World Bank are also subject to monitoring under the CSD. In addition, the 53 elected members of the CSD have been specifically charged with monitoring financial and technical commitments of UN member nations.

Regional groupings have also taken on the challenge of sustainable development and the need for co-ordinated actions amongst their members. For example, at the 1989 Commonwealth Heads of Government meeting in Malaysia, the 49 member countries adopted the Langkawi Declaration on the Environment as a mandate for future action. The heads of the governments of the Commonwealth resolved to 'act collectively and individually' in undertaking a sixteen-point programme of action. This included strengthening efforts by developing countries in forest management, affording support to low-lying and island countries for protection from sea-level rises, promotion of active programmes of environmental education and support measures to improve energy conservation (Commonwealth Secretariat, 1989).

Whilst the increase in international environmental treaties and the creation of new institutions such as the CSD are evidence of the contemporary significance of environmental issues within the conduct of international relations, the real tests of international commitment to the implementation of actions towards sustainable development are intricately linked with other crucial global issues, such as aid, world trade and economic growth.

## Aid and the environment

Foreign aid is defined as any flow of capital to the developing nations which meets two criteria. First, its objective should be non-commercial from the point of view of the donor; second, it should be characterised by interest and repayment terms which are less stringent than those of the commercial world. The concept of foreign aid is that these grants and loans are broadly aimed at transferring resources from wealthy to poor nations on the grounds of development or income redistribution. There is substantial debate over the impact of overseas development assistance (ODA) on the recipient nations. Opinions range from the belief that it is an essential prerequisite for development, supplementing scarce domestic resources, to the view that aid perpetuates neo-colonial dependency relationships which will ensure that recipient nations remain underdeveloped (for key areas of this debate see, for example, Hayter, 1989; Mosley, 1995; Todaro, 1997).

As discussed in Chapter 1, cases and understanding of the detrimental environmental impacts of development assistance rose in the 1970s. In particular, resettlement projects, large dams and road building were seen to have caused widespread damage to environments and local peoples. Into the 1980s, *The Ecologist* magazine, for example, was active in publicising the environmental degradation caused by, for instance, iron ore extraction and highway projects in Amazonia. However, there is growing recognition in the 1990s that partnerships between industrial and developing countries may be indispensable to achieving sound environmental management on a global scale; that is, aid can be (and has been) environmentally damaging, but it is also needed if poverty is to be alleviated and environments are to be conserved. The Labour government elected in Britain in 1997, for example, set sustainable development at the top of its development assistance agenda, and recognised explicitly the interdependence of aid with areas of trade and wider foreign policy, as encapsulated in the Department for International Development Document *Eliminating World Poverty: A Challenge for the 21$^{st}$ Century* (1997):

> Development assistance is an important part of the way in which we can help tackle poverty. But it is not by any means the only aspect of our relationship with developing countries. Both nationally and inter-nationally, there is a complex web of environmental trade, investment, agricultural, political, defence, security and financial issues which affect relations with developing countries. These are driven by a range of policy considerations, all of which affect the development relation-

ships. To have a real impact on poverty we must ensure the maximum consistency between all these different policies as they affect the developing world. Otherwise, there is a risk that they will undermine development, and development assistance will only partly make up for the damage done.

Source: Department for International Development (1997)

Figure 3.2 gives further insight into how the British government is attempting to give human rights concerns a higher profile in foreign policy (including through the publication of an annual report auditing its activities in this respect).

The World Bank (WB) group, which consists of the International Bank for Reconstruction and Development (IBRD), the International Development Association (IDA), the International Finance Corporation (IFC) and the Multilateral Investment Guarantee Agency (MIGA), is the major source of multi-lateral aid for developing countries. In 1996, gross disbursements from IBRD and IDA totalled almost $20 billion (World Bank, 1997a). In addition, for each dollar that the World Bank lends, it can be expected that another two to three dollars will also flow to these projects from other agencies, from private banks and from the recipient governments (Rich, 1994). The rhetoric and actions of the World Bank with regard to the environment are therefore crucial in determining the prospects for sustainable development.

The World Bank (with its 'sister organisation', the International Monetary Fund (IMF)) was established at the Bretton Woods Conference in 1944, and was part of a package aimed at ensuring the reconstruction and development of Europe after the Second World War. In operation, since 1950 the WB has lent monies increasingly to the governments of the developing nations. The IMF is primarily concerned with economic and financial stability and sets and oversees codes of economic conduct on behalf of members. To qualify for WB loans, countries must first be members of the IMF.

In 1993, the World Bank announced its four-fold environmental agenda in recognition of the way in which environmental degradation was threatening the attainment of its objectives, stated as being 'to reduce poverty and promote sustainable development' (World Bank, 1994). Central to the agenda were procedures for 'greening' project lending – that section of WB lending which goes to individual governments for specific projects as identified and designed by the governments of recipient nations in collaboration with WB personnel. Environmental assessment (EA) is now a part of the World Bank project cycle: before projects are approved, they are 'screened' for

**Figure 3.2**  *Assessing Britain's 'ethical foreign policy'*

........................................................................................................

| Country | British initiatives | The regime as assessed by Amnesty International in 1997 |
| --- | --- | --- |
| Algeria | Foreign and Commonwealth Office Minister led EU mission in January to continue political dialogue and to express concern for and sympathy with Algerian people Group of Algerian women MPs visit Britain | Thousands killed by security forces and government-backed militia Hundreds of civilians killed by armed opposition groups Many political detainees held for several years without trial Hundreds of death sentences imposed without trial |
| Bosnia | £1 million support for new government £42,000 in 1997/8 for the Open Broadcast TV Network £3 million to the 'Trust Fund for Police Restructuring in Bosnia' £1.2 million to help exhume mass graves | Thousands of Bosnian Muslim men still 'missing' 250,000 new refugees in 1996 Dozens of prisoners held without trial by all sides; some tortured Prisoners often detained as hostages |
| Jordan | Britain offers help to establish a Family Protection Unit to deal with rape and child abuse cases Continuous support, training and technical assistance to the Unit Support for the country's first Rape Crisis telephone line | Hundreds detained for political reasons 500 arrested after anti-government protests in August 1997 100 prisoners received unfair trials 9 executed in 1996 and 21 sentenced to death |
| Pakistan | Britain funding several projects to improve position of women, including in areas of domestic violence, health, gender discrimination and education Human rights training for police | Torture widespread in 1997 70 alleged deaths Death sentences passed on 35 people, one executed Deliberate and arbitrary killings by armed opposition groups |
| China | Britain maintains dialogue with China aimed at encouraging change Supports practical projects including in civil and political rights and areas of economic and social rights | Mass executions (3,500 in 1997) and over 6,000 death sentences Torture is widespread Thousands of political prisoners remain in prison |

| | Concern about arbitrary detentions, the frequent use of the death penalty and constraints on religious and cultural freedoms | 18,000 Muslim nationalists arrested in Xinjiang between May and July 1997 |
| --- | --- | --- |
| *Indonesia* | Funding for a computer database at the Indonesian Human Rights Commission | 350 political prisoners detained; over half did not receive a fair trial |
| | Contributions to the Legal Aid Foundation with its casework Funding for Indonesians at Oxford University on a human rights law course | Torture of detainees, including juveniles, some resulting in death 'Disappearances' in East Timor; dozens killed by the security forces 26 people on death row |

Note: Amnesty International is a campaigning movement that works to promote (and oppose the abuse of) human rights.

Source: based on Black (1998).

environmental impacts and various levels of investigation are triggered according to the projected severity of these and before the project can proceed to approval.

The seriousness of the World Bank's commitments to environmental reforms has already been tested. For example, in May 1991, the USA (the World Bank's largest contributor) threatened to withhold 25 per cent of its 1992 contribution (approximately $70 million). The focus of attention was the WB's US$5 billion involvement in the damming of the Narmada river in India. This project started in 1987 with the construction of the Sadar Sarovar dam, although it had been at least twenty years in the planning stages. At the time, it was the world's largest hydroelectric and irrigation complex, based on 30 major, 135 medium and 3,000 minor dams to be built over fifty years. It was designed to generate an estimated 500 million megawatts of electricity, irrigate over two million hectares and bring drinking water to thousands of villages.

The Narmada scheme, however, has been shrouded in controversy from the outset. The dams will displace 200,000 people, submerge 2,000 sq. km of fertile land and 1,500 sq. km of prime teak and sal forest, and eliminate historic sites and rare wildlife. The scheme has been referred to as an environmental catastrophe, a technological dinosaur and an example of flagrant social injustice (Schwarz, 1991). Fierce local and international protests against the scheme led to the World Bank taking the unprecedented step of commissioning an independent review of its activities on the scheme. In 1993, the decision was made to withdraw its support for the programme. Although the Indian government is continuing with the scheme, public opposition continues. In 1998, for example, all work was stopped on the Maheshwar dam site, when 10,000 villagers, who were set to lose their homes and lands if construction was completed, engaged in one of the largest peaceful sit-ins in Indian history (Vidal, 1998).

A further element of the World Bank's environmental agenda has been to provide funds for projects which specifically aim at strengthening environmental management. By 1995, almost $10 billion was committed to 137 'environmental' projects, as shown in Figure 3.3. Such projects include funds for research, capacity building, training and monitoring, as well as direct investment in pollution prevention and treatment (which accounted for approximately 60 per cent of this lending by 1995), conservation of biodiversity, integrated river management and establishing national parks.

The World Bank was also integral in establishing (with UNEP and UNDP) the Global Environment Facility (GEF) in 1991. This is a programme of new monies (over and above existing ODA contributions) to assist the least developed nations in tackling

**Figure 3.3**  *World Bank targeted programmes for the environment*

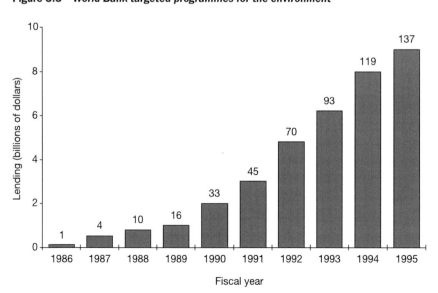

Note: numbers at top of columns represent the total number of active environmental projects.

Source: World Bank (1995).

explicitly global environmental problems, including the limitation and reduction of greenhouse gases, the protection of biodiversity, international water management and energy conservation. The projects approved under GEF to date, by thematic area, are shown in Figure 3.4.

> Praised by some as an important tool for future multilateral co-operation on environmental issues, it is roundly criticised by others for not addressing the key environmental concerns of developing countries. In fact, debate over the extent and nature of GEF reforms has become for some an opportunity to address the reform of development assistance in general.
>
> (WRI, 1994: 229)

In 1994, GEF was extended from its initial pilot phase and further monies committed, with the World Bank acting as trustee, the UNDP providing technical assistance and preparing projects, and UNEP planning to ensure the integrity of GEF projects (Werksman, 1995). Critics focus on the domination of the World Bank and the difficulty therefore of constructing any significantly different model of development (WRI, 1994). For example, membership of GEF was limited to those countries making a minimum contribution of $4 million to the fund and was therefore dominated, 'just like the World Bank's board of executive directors', by developed countries

**Figure 3.4** *GEF investment by sector*

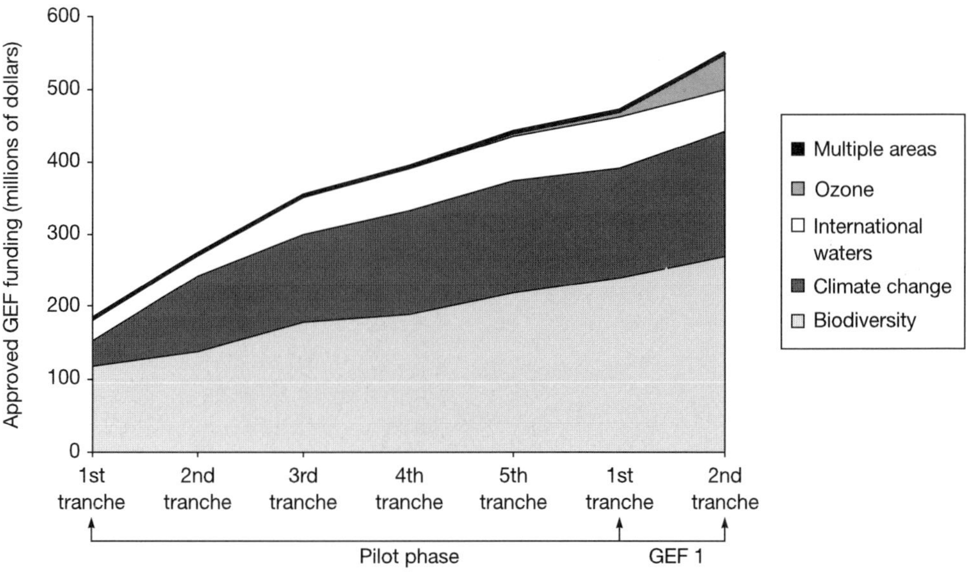

(Werksman, 1995: 282). During the pilot phase, 80 per cent of GEF projects were linked in some way to larger ongoing World Bank projects (Werksman, 1995). To date, therefore, GEF has had limited impact on the reshaping of traditional development lending. Furthermore, the budget ($103 million in 1995 according to the World Bank) is very small in terms of the total financial needs outlined at Rio (WRI, 1994).

As identified in Chapter 1, an increasing proportion of World Bank funds are now directed at broad-based policy reforms (SAPs) rather than specific projects. Once again, this has impacts for the flow of bi-lateral aid as well. Some of the concerns regarding the actual and prospective environmental impacts of policy lending are considered in subsequent sections. The World Bank itself has recognised that its existing procedures and methods for integrating environmental impact assessment into the project cycle are not adequate for designing, monitoring or evaluating such policy interventions. However, 'strategic environmental assessment' is still a relatively novel planning tool in a developed world context and has not, to date, been widely integrated into World Bank activities.

# Trade and the environment

Table 3.1 *The rising importance of trade in the economies of selected countries (trade as a percentage of GDP)*

| Country | 1980 | 1995 |
| --- | --- | --- |
| Bangladesh | 24 | 37 |
| Kenya | 67 | 72 |
| Ghana | 18 | 59 |
| India | 17 | 27 |
| China | 13 | 40 |
| Philippines | 52 | 80 |
| Brazil | 20 | 15 |
| Chile | 50 | 54 |
| United Kingdom | 52 | 57 |
| Sweden | 61 | 77 |
| United States | 21 | 24 |
| Japan | 28 | 17 |

Source: World Bank (1997b).

At both a global and national level, trade is an increasingly important element in development. In recent years, for example, world trade has been growing at more than twice the rate of world output (Buckley, 1996). More of what is produced is now sold in external markets around the world, with services (including tourism and international finance) constituting the most rapidly expanding component of that trade. Globally, the trend has been for exports to form a higher proportion of Gross Domestic Product over time, as shown in Table 3.1. Very recently in fact, international production (rather than exports) has become the dominant mode of servicing foreign markets (UNCTAD, 1997). However, ownership of multi-national companies is concentrated in a small number of more developed countries and exports remain the principal route for the developing nations to secure finances for national development.

Whilst global trade is expanding rapidly, there is evidence that the benefits are far from fairly or widely distributed. For example, people in many countries are experiencing insecurity in employment and deterioration in labour conditions. In September 1998, the Japanese electronics firm, Fujitsu, announced closure of its £5 billion factory in North Tyneside, England, after only six years in production, with the loss of 600 local jobs (*The Guardian*, 1998). This followed the closure in the previous month in the same region of a micro-chip plant owned by the German multi-national firm, Siemens, with the loss of 1,000 jobs after only 15 months. There is also evidence that pressure on the environment is increasing with the expansion of trade world-wide. Although recognition of the environmental impact of trade policies was one of the hallmarks of the WCED report in 1987, such concern is only very slowly being translated into substantial actions, certainly on behalf of the international institution, the World Trade Organisation, charged with setting the rules and resolving disputes in this arena (Potter *et al.*, 1999).

In 1947, the General Agreement on Tariffs and Trade (GATT) was established through the same statesmanship which created the United Nations, the World Bank and the International Monetary Fund. A set of international trading rules were developed for the promotion of future economic stability and development after the era of economic crisis, heavy protectionism, mass unemployment and the Depression (which had all formed part of the backdrop for the Second World War). In 1995, the World Trade Organisation (WTO) replaced GATT, but it has the same principles of free trade and market liberalisation underpinning its rules for international trading. In 1947, GATT had only 23 signatories. There are currently 122 member states of the WTO (China and Russia being the largest non-members). The value of international merchandise trade over the same period has increased from US$57 billion annually to over US$5,000 billion (Buckley, 1996). Although the WTO rulings refer strictly to international trade policy, the agreements made by the organisation have far-reaching implications for economic development, peace and security and environmental protection world-wide.

In continuity with its predecessor, the WTO has phases of greater action, known as 'rounds'. Decision-making within the WTO is based largely on consensus, with each member getting one vote (rather than being weighted, according to economic contribution, for example as within the World Bank). In part, this system of decision-making explains the length of negotiations; the last round of GATT took seven years to complete. Member states of the WTO agree to two fundamental principles: of 'national treatment', under which countries must treat external participants in their economies in the same way as domestic firms, and of the 'most favoured nation', which states that any concession granted by a member to any one trading partner must be extended to all.

As early as 1971, GATT had a Trade and Environment Committee, although, in practice, it never met. The new WTO has twice as many committees and councils as GATT, including a new Committee on Trade and the Environment. Many of the new committees are aimed at providing stronger procedures for dispute resolution and enforcement and for coverage of new areas of trade (outside the traditional merchandise products), particularly in the increasingly important areas of banking, tourism, intellectual property rights and insurance.

In short, the terms of reference for the Committee on Trade and the Environment refer solely to how environmental measures may impact negatively on trade. There is no consideration of how trade liberalisation may aggravate or cause environmental degradation (Potter *et al.*, 1999). For example, it is possible under the articles of the WTO agreement for countries to regulate trade in certain products in order to protect human, animal or plant life or health. However, any such measure must be applied to domestic as well as foreign firms (i.e. be non-discriminatory) and cannot be used as a protectionist device (i.e. must be clearly for conservation ends and not for trade protection). Critically, WTO rulings focus on the product not the processes which are involved in its production. It is possible, therefore, that a country may restrict the importation of a certain good if it will cause environmental damage. What a country cannot do is stop the importation of a good which has caused environmental damage elsewhere during the course of its production: 'the way the import is produced, if it has no effect on the product as such, is not an adequate reason to discriminate against it' (Cairncross, 1995: 227).

If a country wishes to impose environmental and health standards on productive activities and passes environmental legislation (such as landfill or carbon taxes) towards that end, that country does not have a right under the WTO articles to impose those standards on other countries. As a result, it could be argued that such countries risk making their own production uncompetitive in a world market where goods produced under less environmentally friendly conditions will still be traded.

Improving environmental standards may also prove very difficult in practice, when it is considered that large multi-national corporations (MNCs) currently control the majority of world trade (and particularly in the major products of the developing world such as tea, coffee, cotton and forestry). The economic power of such companies in relation to whole countries was noted in Chapter 1. As an example, the MNC Cargill controlled 60 per cent of world cereal trade in 1990 (*The Ecologist*, 1993). In that year, the company had a gross income equal to that of Pakistan, a country of 500 million people (Bryant and Bailey, 1997). Under such circumstances, it is difficult for host countries to impose environmental controls on production. Furthermore, the WTO rulings themselves make no distinction between enterprises in terms of scale of operation or impact. This factor, coupled with the power of the corporate lobby in trade negotiations within the WTO, ensures that the activities of the multi-national corporations remain largely beyond international regulation as well. Fundamentally, MNCs put profit first and hold no allegiance to any particular place, community or environment (UNRISD, 1995). Environmentalists are generally fearful of the prospects of sustainable development through trade liberalisation.

There are, however, a number of sources of pressure on business and industry to take greater account of their environmental impacts, some of which are highlighted in Box H. Currently, for example, many MNCs have developed sophisticated corporate strategies on the environment in response to public pressure and consumer boycotts against their products in the developed world and fears over similar sentiments emerging in the developing nations.

It can be argued that the WTO, more than any other international instrument or treaty, will determine progress towards sustainable environmental policies. International trade is the most significant dimension of global economic activity. To a large extent, therefore, the new WTO will determine patterns and processes of resource exploitation and will have a considerable impact on many of the world's most pressing environmental problems. Whilst trade can encourage economies to make money in environmentally damaging ways, trade barriers can also lead to environmental damage (if, for example, they lead to depressed world prices such that the wasteful use of resources is encouraged). The revenue generated through trade

## Box H

---

### *The greening of companies*

Corporate attitudes to environmental issues have changed significantly in recent years. In short, pressures from government policy, consumer tastes and industry's own perception of its environmental responsibility have combined to move many companies to behave in a more environmentally responsible manner than previously.

The most common tool of national environmental policy has been regulation or 'command and control': governments set standards, such as for minimum levels of dissolved oxygen in river water or for the amounts of nitrous oxide in the air, and then set about enforcing these standards through regional and local public servants. 'A great deal of corporate environmentalism has been driven by regulation' (Cairncross, 1995: 188). However, such regulatory controls have a number of problems. To be effective they require a well-resourced and powerful regulatory infrastructure to 'police' the enforcement of the legislation. Attempts to impose tighter regulations in one country have also encouraged industry to export its hazards elsewhere (see Box G). Furthermore, in setting a ceiling on pollution, there is little incentive for a company to invest in reducing emissions or wastes substantially below that level.

A less widely used strategy is for governments to induce companies to undertake environmental controls through the creation of economic incentives (via the tax system) to reduce pollution. Taxes are an attempt to put a price on pollution which in theory reflects the costs that fall on society. Examples include taxing sulphur by making leaded petrols more expensive in relation to unleaded (which is done in many European countries) and the charging of higher landing fees to noisier aeroplanes, as is done in Japan, Switzerland, the Netherlands and Germany (Cairncross, 1995). Taxes raise revenues for governments without incurring the same costs as enforcement of regulatory controls, but are politically much more contentious: 'The problem with all new taxes is that somebody has to pay for them – and those who perceive themselves as hurt are usually better at campaigning than those who will benefit' (ibid.: 65).

Change in consumer tastes has also been a powerful factor in the greening of companies. The success of The Body Shop (which has 1,516 outlets world-wide) has indicated that whilst people may not be willing to consume less, they will pay more to consume products which are less environmentally damaging or have the greenest effect. Many companies have responded to this demand by finding ways of using the environmental properties of their products as selling points. Governments and private organisations have supported such moves through 'ecolabelling', aimed at assisting consumers in making their choices. However, there are limits to the role of green consumerism in prompting corporate change. For example, it only affects a narrow range of goods – washing powders and domestic cleaning items rather than television sets. There are tremendous problems in establishing the green credentials of a product from its origins through its use to its disposal ('life cycle analysis'). Consumers may also turn away from products if they do not produce the expected results or break down frequently.

Pressure for companies to behave in an environmentally responsible manner without government intervention may also come from the workforce and managers themselves who wish to have an environmental record they can be proud of.

Source: adapted from Cairncross (1995).

may enable countries to purchase the newest, anti-pollution technologies and protect the environment. However, trade encourages traffic of all kinds, which itself is a major cause of environmental degradation such as climate warming and ozone destruction. The minimal effort made, to date, to assess the environmental implications of international trade, the continued dominance within its membership of representatives of the developed world and multi-national companies, and the lack of representation of consumer or environmental groups within the WTO, are all characteristics to be decried in the name of sustainable development.

## International debt and the environment

A further critical global issue which will test the international community's commitment towards sustainable development in action is that of international debt. Foreign borrowing can be highly beneficial, but it also has its costs in terms of the repayment of loans along with the accumulated interest. The challenge of debt servicing and the key features of the 'debt crisis' in the 1980s were highlighted in Chapter 1. In the favourable global economic climate of the mid-1970s (declining real oil prices, low interest rates and buoyant world trade), heavy borrowing enabled the developing countries to achieve relatively high growth rates whilst still being able to service their debts. The benefit at the time to the developed countries was a dampening of the recession owing to increased export demand on the part of the developing world. Clearly, they also benefited from the interest repayments on loans.

> From the onset of the debt crisis in 1982 through 1990, *each and every month*, for 108 months, debtor countries of the South remitted to their creditors in the North an average six billion five hundred million dollars (US$6,500,000,000) in interest payments alone.
>
> (George, 1992: xiv, emphasis added)

It is estimated that flows of funds *into* the developing world, for the same eight-year period, 1982 to 1990 (including all bi-lateral and multi-lateral aid, direct foreign investment by private companies, and trade credits), amounted to less than US$1,000 billion (George, 1992). Furthermore, despite the vast interest repayments made by borrowing nations, by 1990, the debtor countries as a whole were over 60 per cent more in debt than they were in 1982 and the poorest countries suffered most.

The debt burdens of many developing nations have two major implications for the prospects of sustainable development. First, the need to increase short-term productivity puts pressure on countries to overexploit their natural resources. In the long term, this raises the costs of correcting the environmental destruction inflicted now and reduces the potential for sustained development in the future of resources such as agriculture and forestry. George (1992) suggests that the developing countries which deforested the fastest in the 1980s were, in the main, the largest debtors at that time. Although the links are complex, Table 3.2 illustrates a close association between debt and deforestation. Second, the level of government austerity necessitated by debt servicing reduces a government's capacity to deal with environmental protection and rehabilitation: money diverted to servicing debt is unavailable for environmental management (or indeed, wider programmes of poverty alleviation). Figure 3.5 illustrates how spending on health and education has trailed the cost of debt in Tanzania, for example.

Structural adjustment programmes (SAPs) were designed by the World Bank and the International Monetary Fund explicitly to address issues of debt. Whilst some countries under adjustment are showing improvements in conventional macro-economic indicators, there is also some concern that such processes may be at the expense of the environment (Reed, 1996). For example, the emphasis on encouraging export growth has led to cases of the 'mining' of natural resources. Ghana experienced a 75 per cent reduction in the size of its forest area during the course of its economic adjustment in the 1980s (Rich, 1994). SAPs also regularly demand cut-backs in state spending which often include the budgets and staffing of environment departments (Bryant and Bailey, 1997). There is also evidence that the SAPs are serving to widen socio-economic and gender disparities, impoverishing some of the poorest groups in society and compounding environmental degradation (Potter et al., 1999).

In the late 1980s, specific measures known as 'debt for nature swaps' (DNSs) were piloted as a way of recouping a proportion of loan debts and assisting developing countries explicitly to conserve the environment. Since the first projects in Ecuador and Bolivia, many

**Table 3.2  *The relationship between debt and deforestation***

| Countries ranked by amount of debt[a] | | Countries ranked by extent of deforestation in the 1980s[b] | Percentage of original forest already destroyed[c] |
|---|---|---|---|
| 1 | Brazil | Brazil | 23 |
| 2 | Mexico | Indonesia | 30 |
| 3 | *Argentina* | Myanmar | 51 |
| 4 | India | Mexico | 58 |
| 5 | Indonesia | Colombia | 63 |
| 6 | *China* | Thailand | 83 |
| 7 | *South Korea* | Malaysia | 48 |
| 8 | Nigeria | India | 90 |
| 9 | Venezuela | Nigeria | 61 |
| 10 | Philippines | Zaire | 20 |
| 11 | *Algeria* | Papua New Guinea | 15 |
| 12 | Thailand | Vietnam | 77 |
| 13 | *Chile* | Peru | 26 |
| 14 | Peru | Central America | 82 |
| 15 | *Morocco* | Ecuador | 42 |
| 16 | Central America | Philippines | 80 |
| 17 | Malaysia | Côte d'Ivoire | 90 |
| 18 | *Pakistan* | Cameroon | 25 |
| 19 | Colombia | Venezuela | 16 |
| 20 | Côte d'Ivoire | Madagascar | 61 |

[a] World's top debtors in descending order (countries in italics do not have significant forest reserves on a world scale).
[b] Largest deforesters in the 1980s according to Myers (1989).
[c] Percentage of original cover destroyed in country under (b).

Source: George (1992).

countries have worked with donors, environmental groups and banks to establish projects, including the Philippines, Mexico and Cameroon. The DNSs take various forms, but fundamentally involve the lending agency selling a portion of the debt at a discount to another donor (often an NGO), who then offers the debtor country a reprieve from that portion of debt in exchange for a commitment to a particular environmental project in the country. Another incentive to the debtor nations is that these projects can be paid for in local currency rather than foreign exchange.

The first ever 'debt for nature swap' took place in 1987, when Conservation International paid $100,000 for $650,000 of Bolivian

**Figure 3.5** *How education and health spending trails the cost of debt in Tanzania*

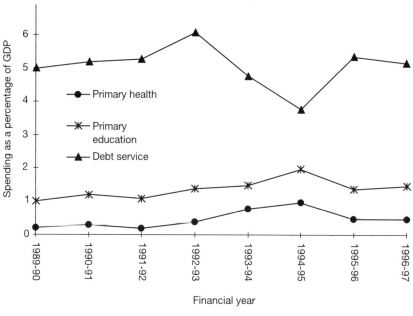

Source: Bush (1998).

debt and forgave it in return for the equivalent of $250,000 in local currency as funds towards the Beni Biosphere Reserve (Marray, 1991). Whilst the impact of these projects on overall levels of indebtedness is small, the local environmental benefits can be significant. However, conservation benefits are long term and it is unlikely that NGOs, multi-lateral banks or individual governments will be able or willing to fund such projects to a level equivalent to the earnings which could accrue from the exploitation of that resource over the short term. This leads to a moral question regarding such an attempt at long-term conservation in the light of unfulfilled short-term needs in indebted countries. To date, DNSs have been applied to a limited range of activities, mainly reserve establishment (Barrow, 1995), arguably reflecting 'Northern' priorities in resource use rather than the needs of local people. Fundamentally, DNSs do little to change the commercial forces which perpetuate environmental degradation.

One of the most recent international proposals for renegotiating the debts of developing countries is the Heavily Indebted Poor Countries (HIPC) initiative agreed at the annual meeting of the IMF and World Bank in 1996. This will be the first time that these multi-lateral institutions will provide relief on debt owed to them. The initiative is supported by the Commonwealth and the G7/8 group of leading industrialised countries. The key objective is to achieve 'debt sustainability', defined as levels of debt which are affordable in relation to revenues without the need for further rescheduling (Department for International Development, 1997). Uganda was the first country to actually receive money under the initiative and Mozambique has recently completed the long qualification process. Under HIPC, countries have to take part in IMF economic reforms for a period of six years before being eligible for relief. There is some concern, however, that the use of solely macro-economic criteria to determine debt sustainability may maximise the debt these countries have to pay rather than providing a platform for real growth, recovery and human development. Several prominent international non-governmental organisations are currently lobbying for assessments of eligibility to be made after deductions of funds for social development and basic human needs have been met, and for measures to ensure that debtor governments spend the money on poverty reduction.

## National action

The state was considered by the Brundtland Commission to have a key role in finding solutions to environmental degradation, in ensuring that the various actors in development, including business and consumers, behaved in the interests of environmental conservation. By 1989, twenty-two governments of the developed world and the European Community had responded to the WCED with details of their progress towards the achievement of sustainable development policies. In particular, the Dutch National Environmental Policy Plan (NEPP) of that year was heralded as a significant landmark, containing 'some of the strongest language seen in an official document on the environment' (Starke, 1990: 39). The Netherlands was the first country to convert the principles of sustainable development identified by Brundtland into concrete steps for action to change both production and consumption (WRI, 1994). The major components of the NEPP are shown in Box I.

By the early 1990s, virtually every country in the world had prepared a national report of some kind on its environment (WRI, 1994). In the developing world, this was generally under the guidance of and through funding from international donors, as shown in Table 3.3. For example, the World Conservation Union (formerly the International Union for the Conservation of Nature and Natural Resources (IUCN)) has assisted over 65 countries to identify internal resource conditions, to determine priorities in environmental action and to

## Box I

---

### *Government of the Netherlands: National Environmental Policy Plan*

In 1989, the Dutch Government declared its National Environmental Policy Plan after three years of consultation and review, and 220 steps for action were prescribed in this ambitious document. Fundamentally, the goal is to produce a structure 'promoting sustainable development through economic incentive, social institutions and self-regulation' (WRI, 1994: 14). Every year, the Dutch environmental ministry issues new four-year plans. Primary targets and responsible actors to date include the following:

**Agriculture:**
- 70 per cent reduction in ammonia emissions (to 30 per cent of 1980 levels by year 2000)
- pesticide use to be cut to half of 1985 levels by the year 2000.

**Transport:**
- Increased emphasis on public transport and bicycles
- Emission ceilings, for nitrous oxide, hydrocarbons, carbon dioxide and noise, set for the years 2000 and 2010
- Removal of tax perk for longer-distance car commuters
- Preparation of 'kilometre reduction plans' by employers
- Encouragement of freight transport by water and rail.

**Physical planning:**
- Aim to secure good public transport access to 'labour intensive' and 'visitor intensive' activities, thereby curbing car use.

**Industry:**
- Reducing the risks of chemical accidents
- Acidifying and eutrophying substances to be cut by 50–75 per cent over 1985 levels by 2000
- Chemical manufacturers to work with government to develop integrated production cycles which reduce waste and pollution.

**Electricity and gas companies:**
- Emission targets set for sulphur dioxide and nitrous oxide
- Energy conservation strongly encouraged, particularly by energy distribution companies.

**Building trade:**
- Doubling of re-use of building and demolition waste by 2000
- Securing 25 per cent improvement in energy conservation
- Substitution of new materials for those with a serious environmental impact
- Initiation of a 'sustainable building' project aimed at incorporating environmental objectives into construction works
- Tightening of insulation standards for houses, offices and other buildings
- Subsidy programme of 70 million Dutch guilders per year for the insulation of existing houses.

**Consumers:**
- By 2000, all used batteries, tin, glass and paper to be collected separately for recycling
- Electricity consumption to be stabilised at 1985 levels
- Growth in passenger kilometres per car to be limited.

Sources: WRI (1994); Ministry of Housing, Physical Planning and Environment (1991).

build the financial and institutional capacities to deal with environmental issues within these countries.

National environmental action plans (NEAPs) are the initial outputs of a process supported by the World Bank. The aim is to assist developing countries in moving beyond environmental reporting and the setting of specific action plans for the environment, towards integrating environmental considerations into a nation's overall economic and social development strategy. Critically, NEAPs are catalysed from within the country itself rather than by the donor community and are drawn up after wide consultation with working groups representing business and industry, public institutions, NGOs and the citizens themselves. The process is also explicitly holistic, recognising the interdependence of sectors including environment, agriculture and health, for example. A high level of government support and advocacy has been identified as critical in the success of the NEAP process, as have strong environmental institutions (which usually need to be established from scratch) and well-motivated and qualified staff (WRI, 1994).

All national governments represented at the Earth Summit committed themselves to the principles of action contained in the Agenda 21 document. In the first three years after Rio, 74 countries submitted to the Commission on Sustainable Development national reports on their activities undertaken to meet the objectives set out in Agenda 21 (Lindner, 1997: 9). The quality of the reports and the processes of planning and consultation on which they are based have been variable. Many countries have also set up national commissions for sustainable development through which NGOs and other interested parties can engage in the Agenda 21 process. Many of the commitments in Agenda 21 need to be enacted at the local level. 'Local Agenda 21' was the framework through which it was envisaged that local governments world-wide would work towards implementation of Agenda 21. Although, to date, LA21 activities have been confined largely to the municipalities of high-income countries, such planning processes are also emerging within countries of the developing world (ICLEI, 1997). Box J highlights progress in the implementation of Agenda 21 processes at the national and local levels in the Philippines and Peru, respectively.

**Table 3.3  Environmental action plans and strategies, 1993**

| Types of action plans and strategies | Africa pub | Africa In prep | Central America and the Caribbean pub | Central America and the Caribbean In prep | South America pub | South America In prep | Asia pub | Asia In prep | Oceania pub | Oceania In prep | Total pub | Total In prep |
|---|---|---|---|---|---|---|---|---|---|---|---|---|
| Canada (CIDA) Environmental Strategies | 1 | 3 | | 1 | 1 | | 3 | | | | 5 | 4 |
| Denmark (DANIDA) Environmental Profiles | 3 | | | | | | 2 | | | | 5 | |
| Netherlands Environmental Profiles | 5 | 2 | | | | | 3 | | | | 8 | 2 |
| United Kingdom/EC/Australia Environmental Synopses | 3 | 28 | | 4 | | 3 | 2 | 2 | | 5 | 5 | 42 |
| United States (US AID) Country Environmental Profiles | 4 | | 11 | | 5 | | 6 | | | | 26 | |
| United States (US AID) Tropical Forests/Biodiversity Assessments | 10 | 3 | 5 | | 3 | | 5 | 1 | 6 | | 29 | 4 |
| CILSS Antidesertification Plans | 6 | | | | | | | | | | 6 | |
| World Bank National Environmental Action Plans | 7 | 11 | | 9 | | 3 | 2 | 14 | | | 9 | 37 |
| IUCN National Conservation Strategies | 14 | 3 | 2 | 7 | | 3 | 9 | 8 | | 2 | 25 | 23 |
| IUCN Conservation of Forest Ecosystem Studies | 8 | | | | | | | | | | 8 | |
| WCMC Biological Diversity Profiles | 13 | | 1 | | 4 | | 5 | | 1 | | 24 | |
| Tropical Forestry Action Programme | 6 | 21 | 7 | 5 | 6 | 3 | 8 | 5 | 2 | 1 | 29 | 35 |
| State-of-the-Environment Reports | 8 | 2 | 2 | 2 | 3 | 2 | 13 | 4 | 1 | | 27 | 10 |
| UNCED National Report | 37 | 6 | 8 | 3 | 7 | | 22 | 5 | 8 | | 82 | 14 |
| Other Policy, Management, and Environmental Studies | 13 | | 3 | | 7 | | 12 | | 1 | | 36 | |
| Total | 138 | 79 | 39 | 31 | 36 | 14 | 92 | 39 | 19 | 8 | 324 | 171 |

Source: WRI (1994).

# Box J

## *Agenda 21 planning*

## The Philippine Agenda 21

Three months after the UNCED conference, President Ramos set up the Philippine Council for Sustainable Development (PCSD) with the main functions of monitoring his government's compliance with official commitments made at Rio and co-ordinating the formulation of the Philippine Agenda 21 (PA21). The Council is composed of representatives from government, people's organisations, NGOs, and the labour and business sectors. The Council received assistance through the UNDP 'Capacity 21' initiative. In 1996, the Philippine Agenda 21 was launched and the PCSD became responsible for overseeing and monitoring its implementation.

Core principles outlined in PA21 include:

- Solidarity, convergence and partnerships between various stakeholders in the state, civil society and the market. In particular, the full participation of marginalised and disadvantaged sectors is required.

- Sustainable development is culturally, morally and spiritually sensitive. It thrives best in an atmosphere of unity and respect for cultural integrity, diversity and pluralism. It is important to nurture the inherent strengths of local and indigenous knowledge, beliefs and practices.

- Gender sensitivity must be the norm. Achieving full equity and equality between women and men and enhancing the participation of women in social development must be a goal.

Source: Department of Environment and Natural Resources (1995).

## Local Agenda 21 in Cajamarca, Peru

In 1993, the mayor of Cajamarca initiated a LA21 effort in what was one of the poorest communities of Peru. The principal challenge was to create a substantially different, decentralised planning body through which all interests could be negotiated and local communities empowered. The city was first divided into twelve neighbourhood councils and the surrounding countryside into 64 'minor populated centres'. Each elected its own mayor and council. An Institutional Consensus Building Committee was established with representation from the province's jurisdictions, NGOs, the private sector and other key groups.

'Theme Boards' were then created to develop action proposals in the areas of education; natural resources and agricultural production; production and employment; cultural heritage and tourism; urban environment; and women's issues, family and population. Each board was charged with developing a strategic plan for its areas of work and training workshops were held to assist them in gathering local inputs.

All the plans were subsequently integrated into a Provincial Sustainable Development Plan, submitted to the provincial council in 1994 and then for public approval through a citizens' referendum. In total, the LA21 process in Cajamarca has raised more than US$21 million for activities, including providing potable water, sanitation, environmental education and rural electrification.

Source: ICLEI (1997).

**Figure 3.6**  *Examples of environmental taxes*

| Tax base | Where taxed |
| --- | --- |
| Solid or hazardous waste generation | Australia, Austria, China, Finland, France, Netherlands, Poland, United States, many municipalities in industrial countries |
| Fresh water use | Australia, Belgium, Denmark, Finland, France, Germany, Ireland, Netherlands, Poland, Turkey, most former Soviet republics |
| Sales of fertilisers or pesticides | Austria, Finland, Norway, Sweden |
| Water pollution | Australia, Belgium, Canada, China, France, Germany, Netherlands, Portugal, Spain, most former Eastern bloc nations |
| Air pollution | China, Denmark, France, Japan, Norway, Portugal, Sweden, most former Eastern bloc nations |
| Production of ozone-depleting chemicals | Australia, Denmark, Singapore, United States |
| Carbon dioxide emissions | Netherlands, Scandinavian countries |
| Motor fuels sales and car ownership | almost all countries |

Source: Brown (1996).
Note: Place lists are not necessarily exhaustive

Clearly, there are a host of specific measures which governments can take to help clean up or prevent further environmental destruction. Figure 3.6 highlights a number of environmental taxes which are now common across many countries, including in the developing world. As seen, many countries now tax car use, for example, through levies on fuels. In France, recent legislation has enabled the authorities in Paris to restrict car use in the city when pollution levels reach certain thresholds.

They effect the controls through allowing only cars with even-numbered registration plates to be driven into the city when pollution levels merit such action (and odd-numbered ones on the next occasion).

These few examples of actions taken by national governments towards sustainable development in their countries illustrate the complex and challenging role of the state in conservation action. A fundamental issue is the tension between the state as protector of the environment and its role as developer. In much of the developing world in the decades following independence, the state took a primary role in leading national and, particularly, industrial developments, including through the creation of large state public utilities, nationalised mining and agricultural enterprises. Indeed, 'the very subject of "development" was built on the idea of the state as the main lever for changing economy and society' (Mackintosh, 1992: 61). Whilst the experience of many of the former Eastern European and Soviet states (and several of the Newly Industrialising Countries) has given insight to the negative environmental impacts of such enterprises in recent years, there is the further issue that many of the most powerful groups in societies (both 'capitalist' and 'socialist') may have built up their position precisely through their control of such (environmentally damaging) activities, particularly in mining and energy generation (Bryant and Bailey, 1997). Some of these actors may be political leaders themselves, causing substantial challenges for the state in fulfilling its stewardship role. Corruption (world-wide) is an underestimated factor in the ability of states to foster sustainable development within their borders.

In the 1990s, there has been much attention paid to the question of how the state could work better, or be reformed, in order to secure more sustainable development patterns. As seen in Chapter 1, introducing the market to state activities is a central tenet of structural adjustment programmes. Even more widely, processes of globalisation, including the operation of MNCs across national boundaries, are in many senses decreasing the capacity of states to act within their territories and over their peoples.

## Non-governmental organisations and sustainable development

NGOs are highly diverse organisations, engaged in various activities and operating at a variety of levels.

> The term non-governmental organisation encompasses all organisations that are neither governmental nor for profit. What is left is a residual category that includes a vast array of organisations, many of

which have little in common. They can be large or small, secular or religious, donors or recipients of grants. Some are designed only to serve their own members; others serve those who need help. Some are concerned only with local issues; others work at the national level, and still others are regional or international in scope.

(WRI, 1992: 216)

The term has different meanings in different contexts. In Latin America, for example, the role of membership organisations is more pronounced than in Asia (Farrington and Lewis, 1993). Indeed, whilst in the English language the term non-governmental organisation is read as politically neutral, when translated into Indonesian, for example, it takes on strong anti-governmental tones which are not conducive to developmental work in what is still a politically repressive country, despite the overthrow in 1998 of President Suharto.

NGOs are not able to sign treaties, pass legislation or set targets for emissions as governments are able to. Historically, however, they have been a strong force in lobbying for such actions to be taken, in modifying governmental activities and in contesting the operation of international institutions such as the World Bank. 'Empowering' such organisations is now promoted by various agencies as the route to an alternative development which may be more sustainable, as well as more democratic and efficient, than previous patterns and processes (Potter *et al.*, 1999).

**Plate 3.1** *NGO–state collaboration in slum upgrading, Delhi, India*

Source: Hamish Main, Staffordshire University.

Perhaps some of the best known NGOs are international. Amnesty International, for example, has over 6,000 groups of volunteers in more than 70 countries (UNDP, 1993). Other organisations like Greenpeace or the World Development Movement, and the World Wide Fund for Nature, tackle issues of truly global concern through lobbying, campaigning and direct action in many countries. Korten (1990) suggests that the balance of activities in which international NGOs have engaged has tended to change over time. He notes how NGOs may form initially around relief and welfare issues (reacting to disaster situations for example), but move subsequently into actions to address the underlying causes of suffering and deprivation, into promoting self-reliant development, and away from direct actions at the village level, towards facilitating development by other organisations.

At the other end of the NGO spectrum are many more numerous, local, 'grassroots' or 'people's' organisations. Particularly in the developing world, groups of people come together for all kinds of reasons to help themselves: by pooling labour, to assist in gaining credit or to enable them to purchase goods in bulk, for example. They may also form community groups in response to the failure of government to provide services such as water or sewerage to low-income housing developments. At times, they are formed in response to the inadequacy and unacceptability of what governments

**Plate 3.2**  *Gold mining in the Philippines*

Source: Lourdes Cooper, University of Brighton.

do, for example, in reaction to political repression or police brutality. It is suggested that there may be 50,000 NGOs in developing countries. In the Philippines alone, there are 18,000 registered NGOs (UNDP, 1993: 86). Clearly, there are problems of actually identifying the precise numbers of such diverse and fluid organisations at this level.

National-level non-governmental organisations have regularly been formed to co-ordinate the activities of local organisations. In the Philippines, for example, there are many umbrella organisations which seek to service and support grassroots membership (people's) organisations in particular areas of activity such as health or land reform. Each people's organisation exists in its own right and is perhaps serviced by more than one NGO. In contrast, in Bangladesh, despite a similar proliferation of NGOs, formal mechanisms for NGO–NGO collaboration have remained weak (Farrington and Lewis, 1993). Alternatively, national non-governmental organisations may have their origins in precisely those 'grassroots', community initiatives, which subsequently spread to form national movements. The Green Belt Movement in Kenya, for example, started in 1977 with a single tree nursery at a primary school and now has mini Green Belts and community nurseries throughout the country.

The rise of the middle classes in many developing regions in the 1980s has been an important factor in the rapid formation of national NGOs. Bryant and Bailey (1997) suggest that the growing size and affluence of the middle classes, their desire for cleaner environments and the advent of more 'democratic' political regimes in many developing countries, have led to a proliferation of NGOs, staffed by professionals and with the objective of lobbying states, businesses and multi-lateral institutions about their environmental policies. These 'Third World Environmental NGOs' are most widespread in South and South-East Asia and Latin America. In contrast, in Africa, the development of the middle class has been arrested through lower levels of economic development generally and by the persistence of authoritarian states which prevent autonomous local activities (Bryant and Bailey, 1997).

Many Northern NGOs are increasingly seeking to work in partnership with Southern NGOs. Indeed, the number of national voluntary organisations in the developed world has also increased dramatically in the 1980s. World Vision in the United States and Action Aid in the United Kingdom are examples of Northern NGOs which have historically raised large amounts of money through private contributions to assist developing countries. Between 1970 and 1990, the amount of grants raised in this sector increased from $1 billion to $5 billion (UNDP, 1993: 88). Such organisations are also serving more widely in recent years as channels for government (ODA) funds: on average, one-third of national NGO funding in industrialised countries comes now from governments, ranging from 10 per cent in the UK to 80 per cent in Belgium and Italy (ibid.).

The environment and the pursuit of sustainable development actions have been primary factors in promoting the growth and importance of NGO activities globally and for stimulating interactions between these organisations and state and/or international institutions of development. For example, in 1997, 47 per cent of World Bank operations involved NGO participation in some capacity, as shown in Table 3.4. In contrast, throughout the 1970s and 1980s, the number of World Bank projects with which NGOs were involved each year was only thirteen (WRI, 1992). Table 3.4 also shows the particular importance of NGO involvement in World Bank projects within the environment and agricultural sectors. Whilst traditionally NGOs have been vociferous opponents of the World Bank, increasingly, they are now involved in the planning and evaluation stages of the project cycle as well as in implementation.

The WCED recognised the key role which NGOs could have in

**Table 3.4**  *Patterns in World Bank–NGO operational collaboration by sector, 1997*

| Sector | No. of projects | % of operations |
|---|---|---|
| Agriculture | 43 | 81 |
| Education | 18 | 56 |
| Electric power and energy | 17 | 18 |
| Environment | 12 | 100 |
| Finance | 14 | 29 |
| Health, population and nutrition | 15 | 60 |
| Industry | 5 | 40 |
| Mining | 2 | 50 |
| Public sector management | 20 | 5 |
| Social sector | 16 | 69 |
| Transportation | 27 | 26 |
| Urban development | 13 | 54 |
| Water supply and sanitation | 13 | 69 |
| Total | 241 | 47 |

Source: World Bank (1997a).

fostering sustainable development based on their proven ability to secure popular participation in decision-making. Experience is suggesting that this can best be secured through processes of planning and action which put people's priorities first rather than those defined by outside actors and agencies (Chambers, 1983). NGOs have traditionally shown greater flexibility and adaptability than larger and more bureaucratic government institutions. Sustainable resource management also therefore depends on a local body through which people's own values and needs can be discussed, planned for and acted upon. In particular, poverty is a symptom of people not having the power to control the resources on which they depend. Sustainable development requires that the poorest sectors of society become agents of their own development ('empowered'). NGOs have often traditionally worked with some of the poorest groups at the grassroots levels and this is a further reason for the attention paid to these institutions in current research and development.

In the following chapters, illustrations will be given of the specific actions of NGOs in rural and urban environments. Although as will be seen, NGOs have certainly been involved in many development actions which are showing signs of being more sustainable than centralised, 'top-down' initiatives, change is needed throughout the hierarchy of institutions discussed in this chapter. Community organisations, for example, need to move beyond actions in their traditional arena of immediate practical concerns, to engage with wider political processes. This may involve challenging local and state power structures and therefore be unpopular. A further challenge for NGOs may be to maintain their accountability to local communities and 'watch-dog' functions as more official aid is channelled through them and they are increasingly required to operate in commercial markets in the delivery of services. Furthermore, it cannot be assumed that all interests are served through 'community development efforts' (Potter *et al.*, 1999); all societies are highly differentiated including by gender, ethnicity and class, for example.

## Conclusion

Many of the actions taken at various levels to promote sustainable development give cause for optimism. It is evident that many institutions of development are transforming what they do in operation, are modifying the ways in which they work with other organisations and are changing their internal structures. Furthermore, new institutions are being created at all levels towards the elimination of poverty and the revitalisation of economic growth and with greater weight given to environmental concerns. It is also clear that further such changes throughout the hierarchy of institutions are required. Indeed, the capacity to generate sustainable development interventions at any level very often depends on actions at other levels. For example, community organisations require an 'enabling' state which allows local democratic processes and the articulation of local voices and needs. The effectiveness of international institutions is bounded fundamentally by the strength of commitments of national governments to develop the rules of behaviour and to ensure compliance within their borders. Such interdependence of actions at different levels confirms the global nature of the challenges and opportunities of sustainable development.

## Discussion questions

* Identify the key debates behind the formation and progress of a key convention, such as on climate change or on trade in endangered species. Consider any tensions which have emerged, for example, between different regions of the world or elements of the scientific community.
* To what extent do you feel that the World Bank is committed to sustainable development?
* Investigate in more depth the arguments for and against free trade, and assess the likely impacts on the environment.

## Further reading

Cairncross, F. (1995) *Green Inc.: A Guide to Business and the Environment*, Earthscan, London.

George, S. (1992) *The Debt Boomerang*, Pluto Press, London.

Mackintosh, M. (1992) 'Questioning the state', in Wuyts, M., Mackintosh, M. and Hewitt, T. (eds.) *Development Policy and Public Action*, Oxford University Press, Oxford, pp. 61–89.

Potter, Robert B., Binns, J.A., Elliott, Jennifer A. and Smith, D. (1999) *Geographies of Development*, Addison Wesley Longman, Harlow.

Reed, D. (ed.) (1996) *Structural Adjustment, the Environment and Sustainable Development*, Earthscan, London.

UNRISD (1995) *States of Disarray: the Social Effects of Globalisation*, UNRISD, Geneva.

# 4 Sustainable rural livelihoods

- The varied and dynamic nature of rural livelihoods in the developing world
- Core characteristics of world agriculture
- The 'marginality' of resource-poor agriculture
- The pre-requisites for sustainable rural development
- Reversing 'normal' professionalism

## Introduction

For the large numbers of people resident in the developing world, their basic needs in terms of both development and conservation are immediate and local; survival in the short term is their primary concern and for this they depend largely on the resources of the surrounding area. For approximately 65 per cent of the people living in the developing world, these needs are also rurally based (UNDP, 1993). Although levels of urbanisation in the developing world are predicted to increase, it is also certain that the absolute numbers resident in rural areas are rising and will continue to do so under projected population increases. Providing sustainable rural livelihoods, not just for the present population but for many billions more, is therefore an urgent endeavour, as these populations will have to be supported by what is often a very fragile and difficult environment.

The importance of rural development in sustainable futures in the developing world stems in part from the numbers of people living in rural areas in these regions. There is, however, much diversity in this respect. On average in Latin America and the Caribbean, for example, only 26 per cent of the total population live in areas classified as rural. In contrast, within countries of sub-Saharan Africa nearly 70 per cent of people are rurally based (World Bank, 1997b). Within these regions, there are also differences between countries, as seen in Table 4.1. The significance of rural development in overall considerations of

**Table 4.1**   *Aspects of the reality of rural living*

| Country | Rural population as percentage of total | People in poverty (%) | | Agriculture as percentage of GDP | Percentage of labour force in agriculture |
|---|---|---|---|---|---|
| | | Rural | Urban | | |
| India | 74 | 49 | 38 | 31 | 64 |
| Zimbabwe | 69 | – | – | 15 | 69 |
| Indonesia | 67 | 16 | 20 | 19 | 57 |
| Uganda | 88 | 33 | 25 | 53 | 93 |
| Nigeria | 62 | – | – | 34 | 43 |
| Argentina | 13 | 20 | 15 | 6 | 12 |
| Brazil | 23 | 66 | 38 | 11 | 23 |
| Sri Lanka | 78 | 36 | 15 | 25 | 49 |
| Peru | 29 | 72 | 52 | 11 | 36 |
| Bangladesh | 83 | 51 | 56 | 30 | 64 |
| Kenya | 74 | – | – | 29 | 80 |
| Malaysia | 48 | 23 | 8 | – | 27 |
| Mexico | 26 | 43 | 23 | 8 | 28 |

Sources: UNDP (1996); World Bank (1997b).

sustainable development stems also from the prevalence of poverty in the rural sector of developing countries, which in the main tends to exceed that in the urban areas as also shown in Table 4.1. The central importance of overcoming poverty if environmental or development goals are to be met in future was detailed in Chapter 2.

Rural areas of the developing world have often been left out of development initiatives in the past. In short, many biases have served to limit the understanding of the needs of rural communities and environments and the implementation of projects and programmes. For example, Chambers (1983) has suggested that limitations on understanding are a product of factors including what he terms 'tarmac bias', referring to the way in which bureaucrats, academics and journalists the world over rarely venture into remote areas; 'person bias', resulting from the tendency to speak only to influential community leaders; and 'dry-season bias', which comes through visiting rural areas when travel is easiest. Michael Lipton (1977) in his classic work on 'urban bias' has argued that the major explanation of the persistence of poverty in developing countries has been the 'anti-rural' development strategies followed therein. In short, his suggestion is that the urban sector has benefited disproportionately from the public allocation of resources in developing countries, in education and other services and through cheap food policies and other public subsidies, for example.

This chapter identifies the key but highly varied characteristics of rural livelihoods in the developing world and the major sources of change within these systems. Such an understanding has over the last decade proved to be essential for supporting actions in rural development that are showing the most promising signs of sustainability.

## Making a living in rural areas

> For adequate and decent livelihoods that are sustainable, much depends on policies which affect agriculture.
>
> (Chambers *et al.*, 1989: xvii)

Aspects of the significance of agriculture in national economies and for individual livelihoods in terms of employment in the developing world are confirmed in Table 4.1. Certainly, in comparison with the general pattern in the 'more developed' regions, agriculture tends to be more prominent in the overall structure of production in the economies of the developing world. However, it can be seen in Table 4.1 that agricultural production has a greater significance currently in some regions and countries than others. Similarly, within rural areas, agriculture can play a varied role in securing livelihoods at the household level. As Rigg (1997) has suggested, 'there is more to rural life than agriculture' (p. 197). Recent research, for example, has estimated that perhaps one-fifth of the rural labour force in developing countries may be engaged in non-farm activities (Chuta and Leidholm, 1990). Generally, however, and in contrast to the situation in the industrialised world, the majority of households of the developing world produce a high proportion of their subsistence requirements (and indeed, the majority of such agricultural production is for subsistence rather than cash purposes).

Diversity is one of the key features of rural livelihood systems, where livelihood is defined as adequate stocks and flows of food and cash to meet basic needs. In the savannah lands of Nigeria, for example, the Hausa and Fulani societies secure their sustenance in close proximity but through very different agricultural systems, based on permanent cultivation and semi-nomadic pastoralism respectively. In contrast, in many parts of South-East Asia, farmers are also regularly industrial workers, travelling perhaps a few kilometres daily or seasonally much

**Plate 4.1** *Income opportunities in rural areas outside agriculture*
      a. Brickmaking in Zimbabwe
      b. Packing flowers, Kenya

(a)

Source: author.

(b)

Source: Hazel Barrett, Coventry University.

further, to take up waged employment, without giving up agriculture altogether. In Latin America, larger numbers of rural people secure a living through agriculture, but as wage labour on plantation estates owned by others.

Dynamism is a further key feature of rural livelihoods in the developing world. Indeed, the capacity to move the emphasis of any particular element within the livelihood system or to introduce new components has been central often to survival itself. Recent research into how poor households cope with food insecurity in times of drought, for example, has highlighted the varied adaptations to changing environmental and social circumstances which households can and do make. The key concern, however, is whether these responses constitute movement towards more secure rural livelihoods or greater vulnerability as considered in Box K.

The concept of entitlements has been useful in understanding the variation which exists between and within rural households in terms of the command they have over food, but also other resources. This concept originated in the work of Amartya Sen in *Poverty and Famine* (1981) and was referred to in Chapter 2 in terms of broadening notions of poverty beyond those which have tended to stress income levels, for example. Access to land is clearly a critical resource for agricultural production, and yet inequalities in landholding in many developing countries are very large, as can be seen in Table 4.2. Furthermore, such inequalities are becoming more entrenched despite decades of land reform programmes (Potter *et al.*, 1999). The basic suggestion within the idea of entitlements is that people differ, not only in terms of the tangible assets which they have (such as land, equipment and stores), but also in terms of their ability to access further resources for livelihoods. Figure 4.2 distinguishes the tangible assets (endowments) from the less tangible claims and access of individuals (entitlements), with particular reference to food resources in this case. A whole host of factors at various levels affect individual entitlements with respect to particular resources (see Young, 1996, for details with regard to food). In the case of access to land resources, women as a group, for example, may be denied access through elements of the legal system and/or the cultural system, which define how, when and by whom land may be held or inherited within a community.

The concept of 'agro-ecosystems' as illustrated in Figure 4.3 captures aspects of both the diversity and dynamism of rural livelihood systems. The model confirms how farming is only one option for securing basic needs for food and cash in rural areas, and that farming itself may be based on a combination of livestock and/or cropping systems. Further diversity stems from the limitless number of factors shaping individual farming and livelihood systems, each farming system involving the manipulation of basic ecological processes (such

# Box K

---

## *Coping with drought: improved security or increased vulnerability?*

If you change a man's way of life, you had better have something of value with which to replace it.

(Kikuyu proverb)

Drought is not a new phenomenon in many areas of the developing world. What is new is the level of suffering with which drought has been associated in the latter part of this century. As Stock (1995) comments with respect to Africa, 'from the late 1960s to the early 1990s, hardly a year has passed without reports of famine in some part' (p. 178). Drought is not a sufficient condition for famine, but it is often the final trigger prompting large-scale starvation and death. Since drought has undoubtedly been an important factor around which rural communities in the developing world have had to adjust their activities for many centuries, research is now focused increasingly on understanding why it is only recently that people, in such large numbers, have been unable to cope with it.

Such research has shown that there are many adjustments that individuals and communities can and do make to reduce the impact of factors such as drought on their food supply. For example, as shown in Figure 4.1, where food production is insufficient, coping strategies may involve a range of actions to secure food through other entitlements or to reduce demand for food within the household. Specific actions include economic strategies such as sales and exchanges in formal markets. However, it is increasingly recognised that non-market transfers such as of gifts of food or loans of cattle may be very important for survival in times of hardship (Adams, 1993). These strategies trade on the social relationships at various levels: between families, through lineage groups or clan membership and amongst neighbours. Assistance may be found through these institutions in terms of gifts of food, loans of cash or even the temporary support of members of one household within another. Such assistance between members of communities is usually based on reciprocity; it may represent repayment of past kindness or a commitment to help in the future.

Such coping strategies amongst communities living in harsh environments have developed over many years and evidence suggests that additional strategies are evolving as the physical, political-economic and social environments change for these people. However, there is much debate as to whether these coping strategies constitute a source of optimism and a basis for future development, or a 'cop out' for development planners (see for example, Davies, 1993).

Watts (1983) argues that peasant households have become more vulnerable to food insecurity with integration into the market economy (starting with colonialism). Specifically, he details how the 'colonial triad' of taxes, cash crops and monetarisation has increased socio-economic inequalities within communities and eroded the 'moral economy' which had provided a degree of security for communities living in harsh environments. The breakdown of traditional coping strategies has left large numbers of people increasingly

**Figure 4.1** *Responses to food deficit*

| Trigger event (production) | Behavioural category (consumption) | Strategy (generic) | Response (specific) |
|---|---|---|---|
| | *Protect consumption* | Purchase grain (market exchange) | – sell non-food crops<br>– use off-farm income<br>– sell assets (e.g. animals)<br>– borrow cash<br>– postpone debt repayment<br>– reduce non-food spending |
| | | Receive grain (non market transfers) | – remittances<br>– charity<br>– begging<br>– food aid |
| Grain production deficit | *Modify consumption* | Reduce consumption (ration) | – smaller portions<br>– fewer meals per day<br>– fewer snack foods |
| | | Diversify consumption (change diet) | – less preferred varieties<br>– wild foods<br>– less nutritious diet (no meat or fish) |
| | | Reduce consumers (change household size) | – wife returns to father<br>– children sent to relatives<br>– male temporary migration<br>– betroth daughter |

Source: Devereux (1993).

vulnerable, not just to natural events, such as climatic drought, but also to crises associated with their incorporation into the market economy, such as the fluctuations in cash crop prices. Whilst one set of values, which had regulated successfully the relationship between people and the environment, has been eroded, another effective set has yet to be incorporated. Famine may itself disrupt precisely the social institutions and mechanisms which gave people a degree of support and ability to cope over time, such as when people move to emergency relief centres, breaking traditional networks of communication.

**Table 4.2** *Inequality in the distribution of landholding in selected countries*

........................................................

| Country | Year | Gini coefficient[a] |
| --- | --- | --- |
| Brazil | 1980 | 0.86 |
| Saudi Arabia | 1983 | 0.83 |
| Kenya | 1981 | 0.77 |
| Chile | 1981 | 0.64 |
| Sri Lanka | 1982 | 0.62 |
| Uganda | 1984 | 0.59 |
| Bangladesh | 1980 | 0.50 |
| Malawi | 1981 | 0.36 |
| Republic of Korea | 1980 | 0.30 |

[a] The Gini coefficient is a measure of inequality which ranges from zero to one. The closer the figure to one, the greater the level of inequality.

Source: UNDP (1993).

as the competition between species or the predation by pests), via agricultural processes of cultivation or pest control, and as led by human goals (which may be set at the individual, community, state or international levels).

Change may come via the full range of environmental, economic, political and social factors across this hierarchy. For example, a decline in demand for a particular handicraft item such as basketry (an economic factor) or a lack of rainfall (an environmental factor) may operate at the local level to prompt change in individual livelihoods. Further up the hierarchy, a new national government policy concerning recommended soil conservation practices may also serve to change local cropping patterns. As with any 'system', change in any one element has implications for the functioning of the system as a whole. The impact may be direct or indirect, large or small, immediate or delayed. Understanding rural livelihood systems within the agro-ecosystem framework, therefore, begins to illuminate the challenges for sustainable rural development.

---

**Figure 4.2** *Concepts of endowment and entitlement in respect to food resources*

........................................................

**Endowment:** Refers to the assets owned and personal capacities which an individual or household can use to establish entitlement to food.

**Entitlement:** The relationships through which an individual or household gains access to food. These can be established through *direct entitlement* (through own production and consumption); *exchange entitlement* (command over food which is achieved by selling labour power in order to buy food); and *trade entitlement* which refers to the sale of produce to buy food.

........................................................

Source: Allen and Thomas (1992).

**Figure 4.3** *The hierarchy of agro-ecosystems*

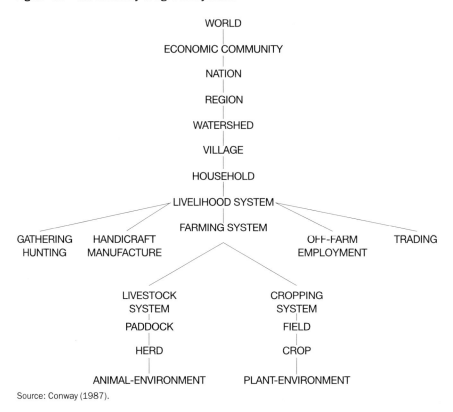

Source: Conway (1987).

## The incorporation of rural areas of the developing world into the global economy

Although many residents of rural areas in the developing world continue to derive their livelihoods as primary agricultural producers, few people or environments now remain outside the workings of the world economy. Agriculture as an economic sector and as an activity has become globalised; it is increasingly dependent on an 'economy and set of regulatory practices that are global in scope and organisation' (Knox and Marston, 1998: 337). So in Chapter 3, for example, it was seen how international institutions organised globally, such as the World Trade Organisation, regulate and dramatically alter agricultural production. In many places (although to varying degrees), the 'farm', which had been the core of agricultural production, is now just one part of an integrated, multi-levelled industrial process including production, processing, marketing and distribution of food and fibre products (Redclift, 1987). The globalisation of agriculture has ensured that changes in the wider global, national or regional economy will also affect agriculture and, conversely, problems in agriculture will affect other economic sectors across these scales.

Figure 4.4 summarises the principal forms (not mutually exclusive) through which developing world agriculture has been incorporated into global markets. The conquest of the New World and the slave trade, for example, enabled the establishment of new plantation systems of production, involving new crops and new labour processes, in countries including Brazil. British settler colonialism in sub-Saharan Africa led to the expropriation of lands from indigenous ownership and brought large areas of agricultural lands under cash crop production for external markets. But it was through taxation and the monetisation of local economies, often initiated during colonial periods, that the most far reaching impact on agriculture and rural livelihoods in the developing world occurred.

Today, land and labour almost ubiquitously have a monetary value in rural areas of the developing world; people have to sell some of what they produce and buy some of their household requirements. Certainly, the relationships between people in rural areas and between people and the environment have changed dramatically through all these forms of incorporation. Through such processes, some people have been able to make profits, expand their operations and move into non-agricultural activities. For many people, however, and particularly small-scale peasant households, the outcome has been

**Figure 4.4** *The major forms of incorporation of agriculture into the world economy*

| Form | Example |
| --- | --- |
| Plantations and estates | Caribbean, Brazil |
| White settlements | Kenya, Rhodesia, South Africa |
| Taxation, rent and other levies in specified crops | Dutch East Indies |
| Taxation, rent and other cash levies | West Africa |
| Taxation, rent and other levies in labour | Dutch East Indies, much of Latin America |
| Creation of new cash needs to purchase imported goods | Burma, Thailand |
| Restriction of land area for traditional agriculture | East and West Africa |

Source: Dixon (1990).

increased hardship and insecurity of rural livelihood, as argued, for example, by Watts (1983) in Box K.

In 1987, the WCED suggested that three distinct types of agriculture could be identified across the globe. Effectively, the categorisation was according to the degree of incorporation into the world economy. Firstly, 'Industrial' agriculture was identified as largely confined to the industrialised countries of the OECD, but also as occurring in specialist enclaves of the developing world. This type of agriculture is highly productive but depends on heavy external inputs such as synthetic fertilisers and chemical insecticides. Trans-national corporations (TNCs) have taken a lead role in agro-industrial expansion. A small number of (often the same) TNCs dominate the supply of the key inputs, for example, as shown in Table 4.3. These corporations have also established networks with research institutions, agricultural colleges and even government ministries and regulatory bodies (*The Ecologist*, 1993). In this way, American TNCs such as Monsanto and Cargill have extended their control over world agriculture into biotechnology developments.

Second, the WCED identified 'green revolution' agriculture encompassing the activities of approximately 2.5 billion people in the countries of the developing world, but most widely in Asia. In areas of reliable rainfall or irrigation technologies and close to markets and sources of inputs, high-yielding varieties of cereals and the associated technological packages required for their production (i.e. characteristics very similar to those of industrial agriculture) have transformed pre-existing agricultural systems. Indeed, the central feature of the incorporation of agriculture into the world economy and its 'modernisation' has been this move to high external input systems of production.

In contrast, the third type of agriculture, identified by the WCED as 'resource poor', supports in excess of a further 2 billion people in the developing world, but has generally been 'forgotten' within agricultural developments. As Pretty (1995) suggests:

> Farming systems in these areas are complex and diverse, agricultural yields are low, and rural livelihoods are often dependent on wild resources as well as agricultural produce. They are remote from markets and infrastructure; they are located on fragile or problem soils; and less likely to be visited by agricultural scientists and extension workers or studied in research institutions.

(Pretty, 1995: 31)

In ecological terms, resource-poor farmers are concentrated in the

**Table 4.3**  *Trans-national control over pesticide supply and seed production for industrial agriculture*

| Corporation | Sales ($ million) | Percentage of global market |
|---|---|---|
| *Top ten pesticide suppliers in 1988* | | |
| Ciba-Geigy (Switzerland) | 2,140 | 10.70 |
| Bayer (Germany) | 2,070 | 10.37 |
| ICI (UK) | 1,960 | 9.8 |
| Rhône-Poulenc (France) | 1,630 | 8.17 |
| Du Pont (USA) | 1,440 | 7.19 |
| Dow Elanco (USA) | 1,420 | 7.11 |
| Monsanto (USA) | 1,380 | 6.89 |
| Hoechst (Germany) | 1,020 | 5.12 |
| BASF (Germany) | 1,000 | 5.00 |
| Shell (Netherlands/UK) | 940 | 4.69 |
| *Total* | *15,000* | *75.04* |
| *Top ten seed corporations in 1988* | | |
| Pioneer Hi-Bred (USA) | 735 | 4.90 |
| Sandoz (Switzerland) | 507 | 3.38 |
| Limagrain (France) | 370 | 2.46 |
| Upjohn (USA) | 280 | 1.87 |
| Aritrois (France) | 257 | 1.71 |
| ICI (UK) | 250 | 1.67 |
| Cargill (USA) | 230 | 1.53 |
| Shell (Netherlands/UK) | 200 | 1.33 |
| Dekalb-Pfizer (USA) | 174 | 1.16 |
| Ciba-Geigy (Switzerland) | 150 | 1.00 |
| *Total* | *3,153* | *21.01* |

Source: Tansey and Worsley (1995).

most fragile areas, such as those prone to flooding and adverse climatic conditions. In economic terms, many such farmers are utilising lands of a quality, and at an intensity of production, at which returns on their labour do not exceed the costs. Many farmers find themselves in a downward spiral of borrowing money or resources to feed themselves or to cover their costs of production in the off-season, only to have to pay these back at unfavourable rates and times in the successive season. Because of their economic position, these farmers lack the financial resources to invest in the capital equipment or inputs necessary to raise production or to implement the land use management techniques appropriate to the physical ecology of the area. Such farmers are also referred to as marginal in a political sense; they are often uninformed, disorganised and outside any formal political system. They may also have very little political power in terms of their participation in decision-making or control over the many structures which influence their daily lives, such as local administration or the operation of markets.

## Towards sustainable agricultural development

The specific challenges for sustainable agricultural development are different for each of these three broad types of agriculture. However, they are interdependent concerns. Fundamentally, past patterns and processes of rural incorporation and agricultural developments have been inequitable, such that perhaps as many as one person in five lives in 'a world where food is plentiful yet it is denied to them' (Conway, 1997: 1). Not surprisingly, hunger and poverty are closely related and it is therefore in the areas of 'resource-poor' agriculture that the major challenges for sustainable development lie. Box L highlights a range of examples where some of the poorest sectors of rural society across the developing world are becoming more impoverished and insecure in their livelihoods under contemporary patterns of 'development'.

Whilst poverty and marginality, by definition, are concentrated in areas of resource-poor agriculture, this is not to deny the specific challenges of the sustained development of industrial agriculture. For example, there is mounting evidence from research stations that previous yield increases under mono-cropping of modern cereals are now slowing. Factors such as chemical toxicity and changing soil carbon–nitrogen ratios within these systems demand further external inputs with a resultant decline in profitability (Pretty, 1995). The environmental impacts of modernised agriculture are also becoming clearer. The World Health Organisation, for example, estimated that pesticide poisoning may affect as many as 25 million agricultural workers in developing countries (1990). Pesticide residues have also appeared in human foods, as shown in Figure 4.5. Furthermore, local diets have changed in areas of 'green revolution' agriculture, such that, even as food supply has increased, people may still suffer deficiencies in certain minerals and vitamins. Figure 4.6 illustrates

## Box L

---

### *The deterioration of rural livelihoods*

The following brief examples illustrate the core challenge for sustainable rural development in the future: the combination of mounting poverty, deteriorating environments and the loss of local control over the basic resources required for agricultural production, particularly land.

In **Malawi**, the deregulation of agricultural markets required under structural adjustment programmes in the early 1980s has led to decreased subsistence security for the estimated 35 per cent of the rural population who are smallholders currently operating less than 0.7 hectares. In particular, for those in remote regions, private trading (their only option with the closure of marketing boards and depots) is unprofitable due to operating costs and the distance from storage facilities. Many people have been forced to sell their labour for food at the cost of not being able to work on their own smallholdings at critical points in the agricultural calendar (Harriss and Crow, 1992).

In **Zimbabwe**, retrenchments from industry are causing people to return to their traditional 'homes' in the Communal Areas in an attempt to make a living in agriculture. The resultant population pressure in these already degraded areas is leading to further subdivision of lands and problems for young people in accessing land for livelihood (Elliott, 1996).

In **China**, the current construction of the Three Gorges dam threatens the displacement of more than a million rural people from their lands and homes (Pearce, 1997).

In the **Sudan**, salinisation within the Gezira irrigation scheme has led to water supply problems and the loss of lands for cultivation, particularly amongst small landholders (Stock, 1995).

In **Thailand**, the very rapid expansion of golf course construction has led to substantial loss of agricultural area and water shortages. In 1994, small-scale farmers were prevented from growing a second rice crop as normal, through government restriction on water supplies and consumption. Despite this, golf courses were able to continue to pump water from reservoirs (Traisawasdichai, 1995).

In **India**, outside the fast-growing regions of the north, unemployment in rural areas has risen with the adoption of the green revolution, owing mainly to the cessation of a million petty tenancies. Per capita food output has been falling in most of the country and one-third of the total agricultural area is now declared to be drought-prone (Patnaik, 1995).

In **Kenya**, on the banks of Lake Victoria, the livelihoods of fishing communities are becoming increasingly compromised by the growth of water hyacinths which inundate the shoreline and prevent the launch of fishing boats. Fertiliser and pesticide use on neighbouring agricultural lands is a major factor in the eutrophication of the lake and the rise of algal blooms. Further loss of control over the resources essential to the livelihoods of local fishing communities is also occurring with the increased costs of boats, nets and labour, in part prompted by the rise of absentee boat owners, many of whom are Nairobi-based business people and politicians (Geheb and Binns, 1997).

In south-western districts of **Uganda**, premature deaths from AIDS have led to a shortage of agricultural labour. There is evidence of households no longer growing the traditional labour-intensive crop of matoke (plantain bananas), but switching to cassava and potatoes which had only been grown as safety stocks in case of famine. The social necessity of spending several days at each funeral further limits the availability of farm labour. Nutritional standards are falling and people now sell food to pay for medicines for AIDS (Dunn, 1994).

In **Guyana,** economic liberalisation has led to 80 per cent of the country's state forests being leased out to logging concerns, largely foreign-owned. In many cases, logging concessions have been given with scant regard for the pre-existing claims and titles of the indigenous Amerindian forest dwellers. Impact assessments have revealed a decline in access to traditional foods, shelter and other forest resources of local communities and even the bulldozing of crops (Colchester, 1994).

how iron density has declined in diets in South Asia, as high-yielding varieties of rice have displaced local fruits, vegetables and legumes that traditionally supplied these essential elements.

Indeed, it is now recognised that the 'green revolution' has had very uneven impacts both spatially and socio-economically, which has caused problems in terms of sustainability. For example, green revolution technologies have focused largely on rice and wheat crops which are not grown by large numbers of farmers in South America or Africa; the production benefits achieved through such developments have been closely correlated with the distribution of irrigation; and the mechanisation of agriculture has generally saved male labour time and increased women's burdens in the various stages of production (for the detail of some of these issues, see Pearson, 1992; Bernstein *et al.*, 1992; Potter *et al.*, 1999). Conway (1997) has asserted that a 'Doubly Green Revolution' is now required, one that is 'even more productive than the first Green Revolution and even more "Green" in terms of conserving natural resources and the environment' (p. 45). There is much debate and concern, however, regarding the biotechnology developments on which further increased production depends, particularly in terms of the predominant role of trans-national commercial interests in research and patenting, which are keen to exploit Northern consumer demands rather than the needs of small farmers in poor countries (WRI, 1994; *New Internationalist*, 1997a).

**Plate 4.2**  *Harnessing scarce water resources for agricultural production in Tunisia*
      **a. Tabia and jessour irrigation**
      **b. Water control in the El Guettar oasis**

(a)

Source: author.

(b)

Source: author.

**Figure 4.5** *Levels of pesticide residue in Indian diets*

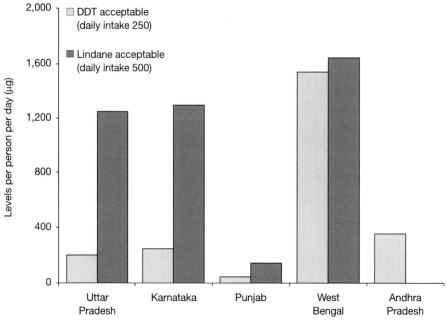

Source: Conway (1997).

**Figure 4.6** *Iron density in diets in South Asia*

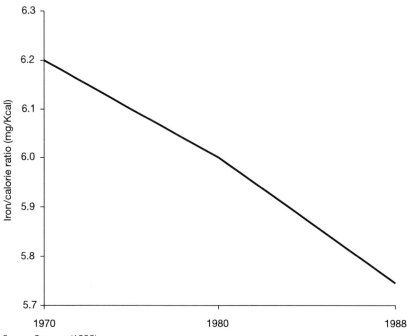

Source: Seymour (1996).

## The importance of 'the local'

In Chapter 1, it was noted that the theoretical necessity of focusing development at the poorest sectors of society had been recognised: the interdependence of environment and development was widely appreciated and a focus on the welfare needs of the poor was required if the goals of either future development or conservation of the environment were to be achieved. The challenge of effecting these commitments in practice is still a relatively new endeavour although many institutions are changing the way they work, as illustrated in Chapter 2. In the rural sector, there are now sufficient examples of successful projects which appear to be sustainable to make suggestions as to the primary characteristics of such projects if they are to be shared elsewhere. The local level is seen to be critical.

Individuals ultimately make the land use decisions upon which sustainable agricultural development depends. They do so, however, within the context of a variety of natural (environmental and ecological) and structural (including the world economy) forces as identified in the hierarchy of agro-ecosystems above and illustrated in Box L. Although it was seen that the range of options in decision-making depend very closely on factors of finance, access to technology and even political power, it is suggested that behaviour is not fully determined, even amongst the poorest farmers:

> Political, social and economic forces do operate; but when they are dissected, sooner or later we come to individual people who are acting, feeling and perceiving . . . all are to some degree capable of changing what they do . . . the sum of small actions make great movements.
>
> (Chambers, 1983: 191/2)

The consistent lesson from recent analyses of apparently sustainable agricultural development is that, indeed, the local level is the key arena for success. For example, Pretty (1995) has documented cases of agriculture delivering economic and environmental benefits across the broad spectrum of the three categories of agriculture identified by the WCED. Whilst the technologies and specific practices may vary, they share three common elements: the use of locally adapted resource-conserving technologies, co-ordinated actions at the local level and supportive external institutions. The types of resource-conserving technologies which are delivering favourable changes to several components of the farming system simultaneously are highlighted in Figure 4.7. Techniques such as intercropping and agro-forestry demand a close understanding of the specifics of local environmental

**Figure 4.7** *Agricultural technologies with high potential sustainability*

........................................................................................................................

| | |
|---|---|
| Intercropping | The growing of two or more crops simultaneously on the same piece of land. Benefits arise because crops exploit different resources, or mutually interact with one another. If one crop is a legume it may provide nutrients for the other. The interactions may also serve to control pests and weeds. |
| Rotations | The growing of two or more crops in sequence on the same piece of land. Benefits are similar to those arising from intercropping. |
| Agro-forestry | A form of intercropping in which annual herbaceous crops are grown interspersed with perennial trees or shrubs. The deeper-rooted trees can often exploit water and nutrients not available to the herbs. The trees may also provide shade and mulch, while the ground cover of herbs reduces weeds and prevents erosion. |
| Sylvo-pasture | Similar to agro-forestry, but combining trees with grassland and other fodder species on which livestock graze. The mixture of browse, grass and herbs often supports mixed livestock. |
| Green manuring | The growing of legumes and other plants in order to fix nitrogen and then incorporating them in the soil for the following crop. Commonly used green manures are *Sesbania* and the fern *Azolla*, which contains nitrogen-fixing, blue-green algae. |
| Conservation tillage | Systems of minimum tillage or no-tillage, in which the seed is placed directly in the soil with little or no preparatory cultivation. This reduces the amount of soil disturbance and so lessens run-off and loss of sediments and nutrients. |
| Biological control | The use of natural enemies, parasites or predators, to control pests. If the pest is exotic these enemies may be imported from the country of origin of the pest; if indigenous, various techniques are used to augment the numbers of the existing natural enemies. |
| Integrated pest management | The use of all appropriate techniques of controlling pests in an integrated manner that enhances rather than destroys natural controls. If pesticides are part of the programme, they are used sparingly and selectively, so as not to interfere with natural enemies. |

........................................................................................................................

Source: Conway (1997).

conditions, substantial management skills and access to information and often significant levels of investment and cost adjustment. Co-ordinated actions at the local level are necessary to ensure that the sustainable development measures of one person or section of the

**Figure 4.8   *Lessons for the achievement of
sustainable rural livelihoods***

...................................................

1  A learning-process approach
2  People's priorities first
3  Secure rights and gains
4  Sustainability through self-help
5  Staff calibre, commitment and continuity.

...................................................

Source: Chambers (1988).

community are not to be compromised by the actions of others. Sustainable agricultural development processes are also most evident where local people are engaged in analysis, decision-making and implementation through supportive and enabling external institutions (both governmental and non-governmental).

Chambers (1988) put forward five major pre-requisites for sustainable rural development on the basis of analysis of apparently successful and sustainable projects in the developing world. These principles, as listed in Figure 4.8, now inform much rural development research and practice in areas within and beyond agriculture. The importance of the local level is again confirmed. However, the more detailed analysis of these five requirements in the following sections indicates that the conditions for sustainability extend also beyond the local level. They include, for example, national systems of tenure which ensure security and encourage long-term investments. Similarly, planning institutions are required at various levels which encourage and give value to community voices.

## A learning-process approach to rural development

Adopting a 'learning-process approach' in rural development planning in the future is in direct contrast to the 'blueprint' approach which has dominated many planning activities in the past. The key features of each approach are identified in Figure 4.9. Fundamentally, projects which have shown signs of being sustainable are continually modified during the course of the project rather than holding to a rigid set of aims and procedures. Changes are made in response to dialogue between all interested parties and the experience gained during the course of the operation of the project. The learning-process approach has implications for how projects are defined, the value of particular kinds of 'expertise' and the systems of communication, for example. As such, the learning-process approach subsumes many of the further 'lessons for success' detailed below.

The core feature of the learning-process approach, the continual modification of 'the project' as it proceeds, is illustrated in the case of Yatenga in Burkina Faso. What started as a forestry project quickly

**Figure 4.9** *The contrasting 'blueprint' and 'learning-process' approaches to rural development*

|  | *Blueprint* | *Learning process* |
|---|---|---|
| Idea originates in | capital city | village |
| First steps | data collection and plan | awareness and action |
| Design | static, by experts | evolving, people involved |
| Supporting organisation | existing, or built top-down | built bottom-up, with lateral spread |
| Main resources | central funds and technicians | local people and their assets |
| Staff training and development | classroom, didactic | field-based learning through action |
| Implementation | rapid, widespread | gradual, local, at people's pace |
| Management focus | spending budgets, completing projects on time | sustained improvement and performance |
| Content of action | standardised | diverse |
| Communication | vertical: orders down, reports up | lateral: mutual learning and sharing experience |
| Leadership | positional, changing | personal, sustained |
| Evaluation | external, intermittent | internal, continuous |
| Error | buried | embraced |
| Effects | dependency creating | empowering |
| Associated with | normal professionalism | new professionalism |

Source: Chambers (1993).

evolved into an initiative focusing on soil and water conservation in response to farmers' priorities. Initially, a system of earth bunds or contour banks was developed and tried, but in practice farmers preferred rock bunds. In conditions of low and erratic rainfall, the farmers' priority was to keep water on their fields rather than to control the movement of water across their plots. Rock bunds are not damaged by runoff and therefore this reduces maintenance; they are permeable, so that crops and therefore yields can be increased; and they encourage the infiltration of water, so raising the effectiveness of the limited rainfall in the area. With such changes, the adoption of bunding increased and, in conjunction with the use of organic manures and improved tillage methods, this led to the reclamation of degraded and abandoned lands in the region, so enabling the expansion of the cultivated area in conditions of high population pressure. Yields increased significantly in the short term, which encouraged more and more farmers to adopt the technology.

## People's priorities first

Putting people's priorities first in the design of projects, rather than those of urban-based professionals, is a further condition for sustainability, but has often not been the case in rural development. Indeed, as already discussed, resource-poor agriculture has generally not been the focus of research and development, yet farmers' and scientists' priorities may differ in a number of core respects, as shown in Figure 4.10. In particular, the question of risk minimisation for

**Figure 4.10**  *Where farmers' priorities might diverge from those of scientists*

|  | Priorities | |
| --- | --- | --- |
|  | *Scientists* | *Resource-poor farmers* |
| Crops | – yield<br>– compatible with machine harvesting<br>– single variety | – flavour<br>– local marketability<br>– multiple variety cropping |
| Cropping systems | – mono-cropping<br>– high external input<br>– high yield | – diverse cropping<br>– low external input<br>– yield less important |
| Management | – maximise production<br>– maximise growth | – minimise risk<br>– livelihood security |
| Use of labour | – minimise labour input | – use all family labour |
| Constraints | – meeting demands of scientific community<br>– project cycles<br>– meeting demands of donors | – meeting traditional obligations<br>– maintaining good community relations |

small-scale farmers has been underestimated in much research station work to date. Chambers (1983, 1993) has been influential in detailing the powerful forces which tend to perpetuate 'first' ('scientist' or 'outsider') priorities, those which start with 'economies not people', with the 'view from the office not the field' and lead to centralised, standard prescriptions for change, rather than the priorities of the 'last' (i.e. farmers). Aid agency bureaucracy, for example, leads to pressure to produce a portfolio of projects quickly and to spend budgets by deadlines, giving little time within the project process to be open to changing conditions and experience from practice or for projects to evolve. Too often, in the past, 'outsiders' have assumed that they knew what poor people wanted when, in practice, the priorities of farmers and of different groups within the local community would vary quite widely.

Accomplishing this requirement to put farmers first requires a substantial number of 'reversals' in 'normal' research and development learning and practice. To understand and learn from indigenous technologies and to build research and extension activities to combine these with 'conventional' science is of paramount importance, as discussed in Box M. This is a central element also for Conway (1997) in defining his 'Doubly Green Revolution':

> Whilst the first Green Revolution took as its starting point the bio-logical challenge inherent in producing new high-yielding food crops and then looked to determine how the benefits could reach the poor, this new revolution has to reverse the chain of logic, *starting with the socio-economic demands of poor households* and then seeking to iden-tify appropriate research priorities. Success will not be achieved either by applying modern science and technology, on the one hand, or by implementing economic and social reform on the other, but *through a combination of these that is innovative and imaginative*.
>
> (Conway, 1997: p. 42, emphases added)

Such conditions for change clearly require the participation of local communities. Although 'participation' has appeared as an objective within many past rural development efforts, regularly this has meant little more than ensuring that such groups were informed about proposed interventions, with no lasting impact on people's lives (Pretty, 1995). 'Self-mobilisation' or, more commonly, 'empowerment' is the term now used to refer to the process of individuals and communities (and typically the 'poorest of the poor') becoming agents of their own development (Craig and Mayo, 1995). Some of the wider forces prompting such shifts in development thinking were considered in chapters 2 and 3, including, for example, the desire for

# Box M

## *The value of indigenous technologies*

> Rural people's knowledge and scientific knowledge are complementary in their strengths and weaknesses. Combined they may achieve what neither would alone.
>
> (Chambers, 1983: 75)

During the 1950s and 1960s, there was tremendous optimism for the role of Western science in raising agricultural production throughout the world, encapsulated in the research and extension activities associated with the 'green revolution'. The locus of research was the experimental station and the challenge was to transfer the new technology to the farmers' fields. When it subsequently became clear that farmers were unable to gain yields on their own farms comparable to those achieved at experimental stations, the 'blame' was passed between 'ignorant farmers' and 'poor extension services'.

It is now appreciated more widely that rarely do farmers fail, through ignorance, to effect land use decisions which will raise productivity or conserve resources. Rather their behaviour is, more regularly, rational in the light of their political-economic, social and environmental circumstances. It is now thought that research conducted at experimental stations has limitations for solving the 'real-life' problems of the farmer (particularly the resource-poor farmer). Scientists have an important role to play in conducting research *about* a problem, for example, how potatoes grow. But for *solving* a problem, such as how to grow potatoes, it is thought that farmers in fact have a lot to teach scientists (Chambers *et al.*, 1989). The problem for research and extension activities, therefore, becomes not how to transfer technology from research station to farmer but how to close the gap between the two so that insights from both can be shared and built upon.

The value of indigenous technology and the benefits of closer links between scientist and farmer can be seen in the example of a technology for potato storage. The success of research into 'diffused light storage' is regularly attributed to the International Potato Centre in Peru (Chambers *et al.*, 1989). However, the basis of the technique was first observed by scientists amongst farmers in Kenya and Nepal. This led to testing and refinement at the research centre and the technique was then passed back to the farmers. However, an investigation into the uptake of the technique in several countries found that only in 2 per cent of 4,000 cases was this technology adopted according to the full recommendations. More regularly, farmers experimented with the technology, selected elements of it and adapted them to suit their own constantly changing circumstances (including financial and social considerations).

However, as more has become known regarding the ways in which farmers learn and experiment, often in very contrasting ways to modern science, it has also become clear that there are differences amongst rural people in terms of their knowledge and power. 'The issue is not just "whose knowledge counts?", but "who knows who has access to what knowledge" and "who can generate new knowledge, and how?"' (Chambers, 1994: xv). Not only, therefore, are there substantial and continued challenges in instilling changes in attitudes, behaviour and methods in the work of institutions and extension agents, but new insights are required into how those who are variously excluded at the local level can be "strengthened in their own observations, experiments and analysis to generate and enhance their own knowledge; how they can better seek, demand, draw down, own and use information; how they can share and spread knowledge among themselves; and how they can influence formal agricultural research priorities."

(Chambers, 1994: xv)

democracy in the developing world and varied pressures for greater accountability in public actions. Empowerment may involve transferring power from one group to another (such as in the case of decentralisation of authority), but more regularly involves creating power through varied activities. These may include, for example, promoting an understanding within local communities of the sources of their lack of power and the actions they may take to overcome this.

Many institutions of development, including government agencies and international and national research centres as well as NGOs, are now working in new ways which 'challenge the traditional top-down process that has characterised so much development work' (Conway, 1997: 199). Participatory research and development methodologies, now variously termed and including 'participatory rural appraisal', 'rapid rural appraisal' and 'participatory action research', have moved understanding forward of the 'complex, diverse and risk-prone environments of resource-poor people' (Scoones and Thompson, 1994: 4). Through the type of research methodologies identified in Figure 4.11, for example, new understanding of rural realities is being gained and is leading to project interventions which are showing greater potential sustainability.

## Secure rights and gains

Ensuring that individual land users and communities have secure rights to resources and the benefits from investments therein is a further condition of sustainable agricultural development based on recent experiences of success. This condition (which is evidently related to the first two) is based on the need to take a long-term view of resource use. When people are sure, for example, that they have the rights to the products from trees that they plant, invest in and manage, they plant many more than they do when there are restrictions on the use or appropriation of such resources. Because of the centrality of agricultural production in rural livelihoods, issues of land ownership and tenure are central to the question of security of rights to resources and benefits. As well as economic opportunity, land is a major

**Figure 4.11** *Methods for participatory learning and action*

| Group and team dynamics methods | Sampling methods | Interviewing and dialogue | Visualisation and diagramming methods |
| --- | --- | --- | --- |
| Team contracts | Transect walks | Semi-structured | Mapping and modelling |
| Team reviews and | Wealth ranking | interviewing | Social maps and wealth |
| discussions | and well-being | Direct | rankings |
| Interview guides | ranking | observation | Transects |
| and checklists | Social maps | Focus groups | Mobility maps |
| Rapid report | Interview maps | Key informants | Seasonal calendars |
| writing | | Ethno-histories | Daily routines and activity |
| Energisers | | and biographies | profiles |
| Work sharing | | Oral histories | Historical profiles |
| (taking part in | | Local stories, | Trend analyses and time lines |
| local activities) | | portraits and | Matrix scoring |
| Villager and shared | | case studies | Preference or pairwise ranking |
| presentations | | | Venn diagrams |
| Process notes and | | | Network diagrams |
| personal diaries | | | Systems diagrams |
| | | | Flow diagrams |
| | | | Pie diagrams |

Source: Pretty (1995).

correlate of social prestige and political power in rural societies of the developing world. Yet substantial inequality in the distribution of landholdings is a widespread and entrenched feature of many developing countries, as seen in Table 4.2. Furthermore, many people (particularly in Asia) are engaged in agricultural production under systems of tenure including share-cropping arrangements, which deliver the major benefits to the landlord rather than the tenant (see for example, Bernstein *et al.*, 1992).

It has been suggested that there are 'two inseparable starting points in any fight for sustainable agriculture: the rights of women and land reform' (Middleton *et al.*, 1993: 124). The limitations of land reform programmes have been briefly alluded to (for a fuller discussion, see Dixon, 1990, or Potter *et al.*, 1999). Box N highlights the fact that women are regularly now suggested as the key to sustainable agricultural development, yet many projects to date do not go far enough towards securing rights to resources for women or challenging

# Box N

---

## *Women and the environment*

> The achievement of sustainable development is inextricably bound up with the establishment of women's equality.
>
> (WRI, 1994: 43)

Throughout the 1980s, it was widely argued that women were the key to sustainable development. Building on work in the 1950s and 1960s which had focused on women's role in the domestic sphere and on bringing women into development through programmes addressing such 'women's areas' as family planning and nutrition, a better understanding of their additional roles in agricultural production and wider reproductive activities, such as fuelwood and water collection and management, developed. Momsen (1991, in this series) details these activities further and Figure 4.12 summarises how women have a 'substantial interest' (Braidotti *et al.*, 1994) in environmental resources.

**Figure 4.12   *Women's 'substantial interest' in the environment***

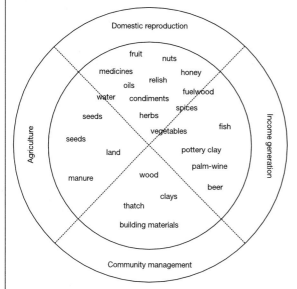

Source: Barrett and Browne (1995).

Through the gendered division of labour (where, for example, women generally take responsibility for caring for children, the active working population and the elderly, and where roles in agricultural production tend to be distinguished by sex), environmental degradation was seen to affect women most. Women tended to take greater responsibility for maintaining communal services such as schools and health posts, and generally often provided the 'glue' between the elements and activities of the community, organising weddings and the like. Women were widely portrayed as 'privileged environmental managers' (Braidotti *et al.*, 1994) because of their close dependence on nature for their subsistence needs and their knowledge of local environments accumulated in the process of their daily activities.

Ecofeminists took these notions further to suggest that women have a natural affinity with nature aligned to their child-bearing qualities, that men do not have. Authors such as Vandana Shiva (1989) have contrasted what they consider women's essential features of caring and empathy with the controlling, manipulative position of men, to argue that women may be the key to new and more sustainable ways of living and social relations, if these 'feminine principles' can be sufficiently recovered.

Certainly, some of the most successful examples of sustainable development to date have been built on women's initiatives. For example, the Chipko movement in India started with a small group of women presenting non-violent resistance to the contract felling of trees in their local area. The principles and practices of this original group have since spread to hundreds of local autonomous initiatives for the protection of forests. Subsequently, many projects have worked with women, building on their existing roles as managers of natural resources as detailed in Figure 4.13. Unfortunately, these have often assumed that women's time is free and infinitely elastic such that the outcome has often been an increased burden on women in rural development.

**Figure 4.13**   *Women organising to manage environmental resources*

........................................................................................................................

\* Women on the Yatenga plateau in Burkina Faso became frustrated at men's inactivity concerning building a dam to capture rains. Women were unwilling to carry water over great distances any longer, so organised to build the dam themselves.

\* In Khirakot village in Uttar Pradesh in northern India, local women took responsibility for managing a community woodlot. When it was threatened by a mining interest, they protested in court and got the mines closed.

\* In Andhra Pradesh, India, village-level women's organisations developed the idea of collectively leasing and managing lands which had become degraded. Although many banks declined to assist the women, they eventually found loans through the Deccan Development Society and have returned over 280 hectares to production through projects which now involve 400 women in 20 villages.

\* In Katheka community in Kenya, twelve, mostly female, self-help groups have formed to collectively construct extensive terraces and check dams to control erosion and enable production in a densely populated, arid land.

........................................................................................................................

Sources : WRI (1992; 1994).

Current work within the 'gender and development' (GAD) school of thought now considers that it is not sufficient to work only with women or to assume that women's relationships with the environment are undifferentiated. For example, gender divisions of labour are not uniform or static (Jackson, 1995). It is now more widely understood that the relations between men and women may vary with class or other social divisions and are continually 'struggled over'. For example, Barrett and Browne (1995) have suggested that 'men help their womenfolk much more than researchers in the 1980s assumed' (p. 32). Gender divisions of labour are also not equivalent to gender divisions of responsibility or control over income. Men often dominate control over the sale and disposal of crops, for example, although these may be produced largely through women's labour. It is evident that such factors filter both the impact of environmental degradation on women and men, and the operation of incentives towards participation in conservation projects.

As such, women may not necessarily be the source of solutions to the environmental crisis. Groups of women as well as men may act in environmentally damaging ways and women's apparent 'closeness' to nature may stem more from their poverty, their lack of opportunities outside the household and the social relations in society, than their biological makeup as prioritised by the ecofeminists. Sustainable development, therefore, depends on removing women's subordination and oppression as well as their poverty, not simply 'grafting on women as a group' or 'lumping' women's varied interests together (Middleton *et al.*, 1993).

fundamental gender inequalities. As an illustration of the extent of the challenge, it is estimated that 80 per cent of food in sub-Saharan Africa is produced by women (Middleton *et al.*, 1993: 124). In Asia, the figure is between 50 and 60 per cent, and it is approximately 30 per cent in North Africa, the Middle East and Latin America. Yet, women own less than 2 per cent of the world's total agricultural area (*New Internationalist*, 1997).

## Sustainability through self-help

In addition, achieving secure rights and gains is essential to ensuring people's perceived self-interest in project development and implementation, which has also been found to be strongly linked to sustainability. It is suggested that, except in the case of deeply impoverished peoples, participation in change should be entirely voluntary and without any form of inducement or subsidy. It appears from experience that when people participate for the sole reason that they have seen success achieved and have become enthusiastic enough to work towards achieving it for themselves, projects tend to be more relevant, to spread more quickly and encourage innovation on the part of the people. A 'self-help' approach aids the sustainability of not only the existing project but also those of the future. However, such voluntarism and action are at the heart of the concept and process of empowerment, which is not straightforward in practice, as discussed above and in Box N. For example, empowerment of particular groups often demands very difficult and complex changes, for example in local politics and gender relations, to ensure that projects do not constitute a further burden on (particularly women's) time.

## The role of 'outsiders'

The final recommendation for sustainable agricultural development refers to the inputs of external agencies. Essentially, good conservation and development solutions encompassing the pre-requisites discussed here in total will not be facilitated by insensitive staff on short-term contracts, for example. To date, staff with the required 'sensitivity, insight and competence' have tended to be recruited by NGOs, where the 'reversals of normal values are often most at home' (Chambers, 1988: 13). It has been more difficult for government field staff to have the close relationship with rural people or freedom of action needed to work in the ways described. However, NGOs themselves are subject to many forces of change which may alter their capacity to work according to these required values. These include the pressure identified in Chapter 3 to compete in a market environment in a context of the increased channelling of development assistance and internal government funds through such organisations. Budgets for salaries of government field staff and for their travel within often remote rural areas are regularly amongst the first to be cut in the austerity programmes required under structural adjustment.

## Conclusion

For too long, the debates about both the environment and development have been dominated by the interests and values of the rich rather than the poor, men rather than women, and urban rather than rural. The results of this emphasis have been seen in this chapter: deteriorating environments, rising poverty and the increasing concentration of resource-poor farmers in some of the most ecologically fragile areas of the world. Reversing these priorities is an essential precondition for sustainable rural development in the future and some of the achievements in this direction have been detailed.

It is evident that whilst individual farmers ultimately make many of the land use decisions in rural areas, ensuring that these are sustainable depends on many wider factors of the international economy and priorities of the state, for example. The continued high dependence of many developing economies on agricultural commodities for exports brings a level of vulnerability to fluctuations in world markets which it could be argued is being accentuated by the encouragement within SAPs to further expand export commodities. However, the impact of such forces on the prospects for sustainable

(a)

Source: author.

(b)

Source: author.

Source: Becky
Elmhirst, University
of Brighton
(c)

Source: Hazel
Barrett, Coventry
University
(d)

**Plate 4.3** *Women in environmental management*

    a. Water collection, Zimbabwe

    b. Fuelwood collection, Zimbabwe

    c. Organising the community: a Lampungese wedding

    d. Preparing fields for agriculture, The Gambia

development in specific regions of the developing world will to some extent depend on national policies, including those in agriculture. Although government policy is a product of a very complex set of factors, key issues in determining the prospects for sustainable agricultural development will include the financial commitment to export or food-crop production, to large- versus small-scale farming and to favoured or resource-poor farmers. Although no single form or scale of production is inherently more or less sustainable than another, past emphasis on the large-scale production of export crops in the more favoured locations of the developing world has been shown to be unsustainable in terms of not only the practices employed but also the people it excluded. A major difficulty at all levels is that the problems of natural resource management within agricultural production have not, to date, been considered adequately.

Making the rural poor the starting point in research and development has been seen to have benefits for the environment and development. The long-term benefits of providing secure rural livelihoods for the rural poor will also fall to the wider communities: rural insecurity is a major factor in maintaining high fertility rates and in prompting rural to urban migration. Thus, improving rural livelihood security will help to relieve population pressure on resources (a central pre-requisite for sustainable development in the future) as well as to slow the demand for employment and shelter in urban areas, which are currently two key demands on the limited financial resources for urban planning in the developing world.

## Discussion questions

* Summarise the characteristics of the majority of farmers in the developing world today.
* Identify some of the possible barriers to the 'farmer first' approach to research and development.
* Compare and contrast the key features of the Green Revolution of the 1970s and the 'Doubly Green Revolution' now being proposed by authors such as Conway.
* Using specific case study examples, consider the relative importance of 'natural' and 'structural' forces determining the prospects of sustainable rural development.

# Further reading

Conroy, C. and Litvinoff, C. (1988) *The Greening of Aid: Sustainable Livelihoods in Practice*, Earthscan, London.

Conway, G.R. (1997) *The Doubly Green Revolution: Food for All in the 21st Century*, Penguin, London.

Pretty, J. N. (1995) *Regenerating Agriculture: Policies and Practices for Sustainability and Self-reliance*, Earthscan, London.

Scoones, I. and Thompson, J. (eds) (1994) *Beyond Farmer First: Rural People's Knowledge, Agricultural Research and Extension Practice*, Intermediate Publications, London.

Visvanathan, N. (ed.) (1997) *The Women, Gender and Development Reader*, Earthscan, London.

# 5 Sustainable urban livelihoods

- Jobless growth in cities of the developing world
- The nature of the Brown Agenda
- Ecological footprints and the regional impacts of cities
- The capacity and effectiveness of city authorities as core conditions for sustainable urban development
- Enabling social organisation at the local level

## Introduction

The proportion of people living in urban areas of the globe is increasing, and particularly in the developing world. Whilst in 1800, only 3 per cent of the total world population lived in towns and cities, it is estimated that by the year 2000, this figure will have risen to over 50 per cent (United Nations, 1989). Although a greater proportion of the population of the developed world currently live in cities, as seen in Figure 5.1 (approximately 73 per cent or 900 million people), the total size of the urban population is larger in the developing world, at

**Figure 5.1** *The proportion of the population expected to be living in urban areas*

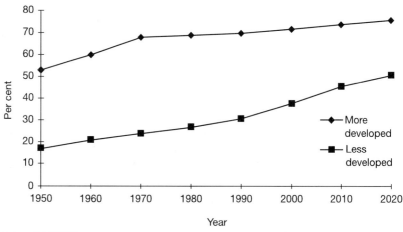

Year

Source: UN (1989).

**Figure 5.2** *Projected population living in urban areas*

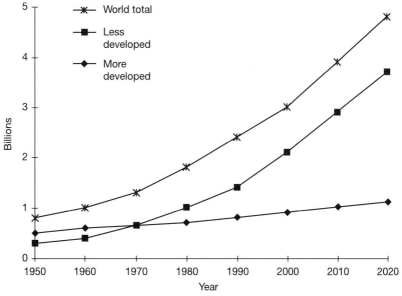

Source: UN (1989).

around 1,400 million people, as seen in Figure 5.2 (Devas and Rakodi, 1993). In addition 93 per cent of the predicted urban growth to the year 2020 will occur in the developing world.

In 1987, the World Commission on Environment and Development suggested that the urban challenge lay 'firmly in the developing countries' (WCED, 1987: 237), due in the main to the unprecedented growth rates, but also to the challenge of meeting the current needs of an expanding urban poor. In that year, for example, the World Bank had estimated that approximately one-quarter of the developing world's absolute poor were living in urban areas (World Bank, 1990a). By the turn of the century, this figure is expected to be nearer 50 per cent, as highlighted in Chapter 2.

Cities are central to attempts at meeting the goals of sustainable development in the sense that this is where the majority of the world's population will soon be located, with all the associated physical demands (such as for food and shelter) and the political, social and cultural requirements associated with the adoption of urban values. In addition, city-based production currently accounts for the majority of resource consumption and waste generation world-wide (WRI, 1996). Throughout history, cities have been a driving force in development processes and, as cities grow, productive activities tend to concentrate in urban centres. For example, an estimated 80 per cent of GDP growth in the developing world in the 1990s originates in cities and towns (Bartone *et al.*, 1994). Wealthier cities, and higher income groups within urban areas, consume the highest levels of resources and contribute disproportionately to waste generation (WRI, 1996).

There are substantial challenges for *all* cities in managing the environmental implications of economic growth, in meeting the needs of their residents and for protecting the environmental resources on which they depend into the future. The focus of this chapter is the particular challenges of cities in the developing world, where it will be seen that the unprecedented rates of urban growth and industrialisation in combination with poverty create distinct and immediate environmental problems which to a large extent are not key concerns in wealthier cities. Figure 5.3 depicts a general characterisation of how environmental problems and the severity of their impacts differ within cities at various levels of income. The 'pollution' of urban poverty that arises from inadequate water supplies, sanitation, drainage and solid waste collection is seen to be the most immediate problem of cities in the developing world. These issues have been termed the 'Brown Agenda'. In wealthier cities, the key challenges for action lie in reducing excessive consumption of natural resources and the burden of wastes on the global environment (WRI, 1996). This 'Green Agenda', encompassing the depletion of water and forest resources, for example, has tended to receive greater international attention, because of the relation to issues of global environmental change such as climate warming.

However, such typologies or distinctions should not distort the common, global challenges of sustainable urban development. Whilst the Brown Agenda is the priority for low-income countries, actions are also needed in the cities of the developing world to promote the efficient use of resources and the minimisation of waste, if they are to prosper in future without the ecological impact of past developments as currently evidenced in 'first world' cities. In addition, as emphasised throughout earlier chapters, processes of globalisation are producing a far more integrated and interdependent world economy into the 1990s. Cities across the globe are experiencing change, not solely in terms of their size, but also in respect of the activities they host and the function they play in the world's economic, trading and political systems (Hamnett, 1995). This chapter details the primary characteristics of these processes of change and patterns of urban development in the developing world in order to understand more

**Figure 5.3** *Economic–environmental typology of cities*

| *Urban environmental problems* | *Lower-income countries (<$650/cap)* | *Lower-middle-income countries ($660–2,500/cap)* | *Upper-middle-income countries ($2,500–6,500/cap)* | *Upper-income countries (>$6,500/cap)* |
|---|---|---|---|---|
| *Access to basic services* | | | | |
| ● Water supply and sanitation | Low coverage and poor quality, especially for urban poor | Low access for urban poor | Generally acceptable water supply, reasonable sewerage | Good; concern with trace substances |
| ● Drainage | Low coverage: frequent flooding | Inadequate; frequent flooding | Reasonable | Good |
| ● Solid waste collection | Low coverage, especially for urban poor | Inadequate | Reasonable | Good |
| *Pollution* | | | | |
| ● Water pollution | Problems from inadequate sanitation and raw domestic sewage | Severe problems from untreated municipal discharges | Severe problems from poorly treated municipal and industrial discharges | High levels of treatment; concern with amenity values and toxic substances |
| ● Air pollution | Severe problems in some cities using soft coal; indoor exposure for poor | Severe problems in many cities from soft coal use and/or vehicle emissions | Severe problems in many cities from soft coal use and/or vehicle emissions | Problems in some cities from vehicle emissions; health priority |
| ● Solid waste disposal | Open dumping, mixed wastes | Mostly uncontrolled landfills, mixed wastes | Semi-controlled landfills | Controlled landfills, incineration, resource recovery |
| ● Hazardous waste management | Non-existent capacity | Severe problems, little capacity | Severe problems, growing capacity | Moving from remediation to prevention |

| *Resource losses* | | | | |
|---|---|---|---|---|
| ● Land management | Uncontrolled land development and use; pressure from squatter settlements | Ineffective land use controls | Some environmental zoning practised | Environmental zoning commonplace |

| *Environmental hazards* | | | | |
|---|---|---|---|---|
| ● Natural and man-made hazards | Recurrent disasters with severe damage and loss of life | Recurrent disasters with damage and loss of life | High risk from industrial disasters | Good emergency response capacity |

Source: Bartone *et al.* (1994).

fully the specific nature of the challenges and opportunities of sustainable development in this sector.

## Urban change in the developing world

## Patterns

Whilst the general trend across the developing world, as seen in Figures 5.1 and 5.2, is for increasing levels of urbanisation, there are significant differences between regions and countries in the patterns of change. For example, it can be seen in Table 5.1 that the highest growth rates to date have been in Africa which is also where the most rapid change in the near future is predicted to occur. However, it is in South and South-East Asia that the largest numbers of people currently reside in urban areas and where the greatest future expansion in terms of additional urban residents will occur. Countries such as India have very large urban (as well as total) populations, for example. Indian cities such as Calcutta and Bombay are amongst the largest centres in Asia (as seen in Figure 5.4) and indeed the world (see Figure 5.5). It should be noted, however, that for some of the largest cities in the developing world, growth rates during the 1980s were significantly slower than during the 1960s and 1970s (UNCHS, 1996).

In many developing countries, a high proportion of the urban population is concentrated in one or two major cities. This pattern is more established in the countries of Latin America than in any other developing region (Hardoy and Satterthwaite, 1989). By 1985, for example, Mexico City, São Paulo, Buenos Aires and Rio de Janeiro all

**Table 5.1  *Urban population change by region, 1970–2000***

| | Urban population, 1985 (millions) | Urban population growth rate, 1970–85 (per cent) | Projected urban population, 2000 (millions) | Projected urban population growth rate, 1985–2000 (per cent) |
|---|---|---|---|---|
| China | 219 | 1.8 | 322 | 2.6 |
| East Asia | 46 | 4.4 | 68 | 2.6 |
| South and South-East Asia | 377 | 4.1 | 694 | 4.2 |
| West Asia | 63 | 4.6 | 109 | 3.7 |
| Latin America | 279 | 3.6 | 417 | 2.7 |
| Africa | 174 | 5.0 | 361 | 5.0 |
| Pacific | 1 | 4.2 | 2 | 4.7 |
| *Total* | 1,159 | 3.7 | 1,972 | 3.6 |

Source:  Devas and Rakodi (1993).

had populations in excess of 10 million. In contrast, there were no cities of this size in the whole African continent by this date (although Cairo came close). In short, Africa's contemporary rapid urban growth rates are occurring over a relatively small base and are more widely distributed across many smaller and intermediate urban centres. Indeed, the 'explosive growth' of mega-cities in the developing world which was predicted in the early 1980s has not generally been realised. Although there is a growing number of urban centres of unprecedented size, something under 5 per cent of the global population live in mega-cities (UNCHS, 1996). New kinds of urban systems are also developing world-wide which include networks of very dynamic, although smaller cities (see Potter *et al.*, 1999). In short, the actual and predicted patterns of urban change in the developing world have been the subject of much analysis and debate and have proved to be highly varied even within nations and not always as expected by researchers or planners.

## Processes

The key processes of urban change in the developing world are certainly without historical precedent. In nineteenth-century Europe, people migrated to the towns and cities in search of employment and economic advancement. The industrial activities located in those

# Figure 5.4 The largest urban centres in Asia

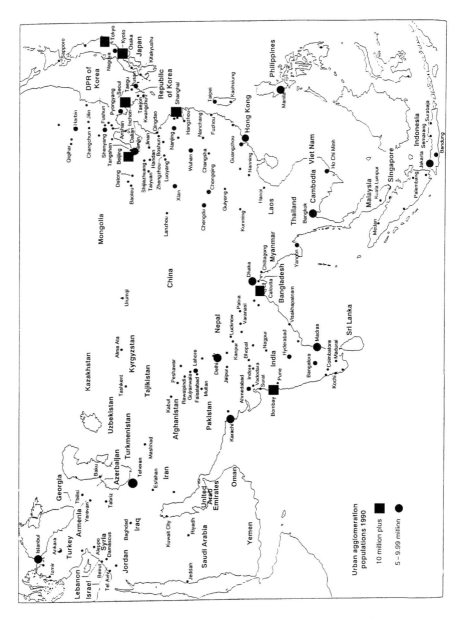

Urban agglomeration
populations 1990

■ 10 million plus

● 5 – 9.99 million

Source: UNCHS (1996).

**Figure 5.5** *The world's largest urban agglomerations in 1990*

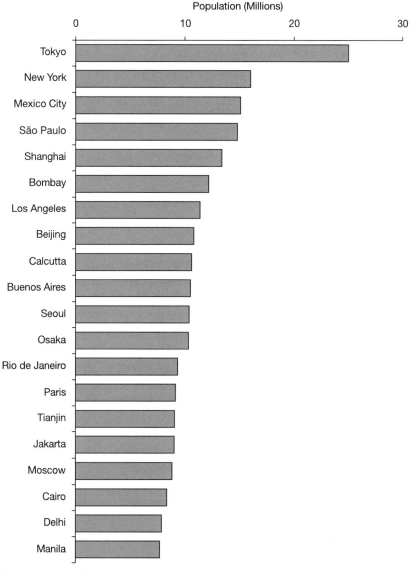

Source: UNCHS (1996).

areas depended on this process of migration to raise output and generate wealth. Urbanisation, industrialisation and 'modernisation' (the adoption of urban values) were processes which occurred simultaneously in the cities of Europe and were mutually reinforcing. This has not been the case in the developing world. Table 5.2 highlights the cases of a number of Latin American countries in the 1960s (a period of relatively rapid industrial development), where it is

**Table 5.2**  *Industrialisation and employment in selected Latin American countries, 1963–9*

| Country | Manufacturing annual output growth (%) | Manufacturing employment growth (%) |
|---|---|---|
| Brazil | 6.5 | 1.1 |
| Colombia | 5.9 | 2.8 |
| Costa Rica | 8.9 | 2.8 |
| Dominican Republic | 1.7 | –3.3 |
| Ecuador | 11.4 | 6.0 |
| Panama | 12.9 | 7.4 |

Source: Todaro (1997).

seen that employment growth lagged substantially behind that in manufacturing output. Such 'jobless growth' continued to be a feature of urban change in the developing world into the 1990s.

Few of the urban poor can afford to be unemployed for any length of time. Many, in fact, will be under-employed; either they are working less than they would like or are doing so at such low rates of production that their labour could be withdrawn with very little impact on overall output. In recent years, structural adjustment programmes have also led to contraction in formal sector employment opportunities in the cities of the developing world, through the loss of jobs in the public sector and the denationalisation of industries, for example. In response to a lack of employment opportunities within this 'formal' sector, many urban residents in the developing world look to a wide variety of both legitimate and illegitimate income opportunities available within the 'informal' economy, the term used to refer commonly to small-scale, unregulated, semi-legal economic activities which often rely on indigenous resources, family labour and traditional technology. Whilst it is now appreciated that the two sectors are not wholly distinct (see Drakakis-Smith, 1987), Table 5.3 shows the estimated share of urban labour force in the informal sector for a number of cities in the developing world. Clearly, in urban areas employment is critical to securing a livelihood and avoiding impoverishment and for sustained development. Todaro (1997) has suggested that one of the most 'obvious failures' of the development process over the past few decades has been 'the failure of modern urban industries to generate a significant number of employment opportunities' (p. 247).

## Poor people in poor environments

Table 4.1 illustrated that the proportion of people estimated to be below poverty lines is generally higher in rural than urban areas of the developing world. However, the absolute numbers of people living at

**Plate 5.1** *Urban informal income opportunities in Harare, Zimbabwe*

    a. Welding

    b. Fuelwood sales

    c. Artefacts

(a)

(b)

(c)

Source: author.

**Table 5.3** *Estimated share of urban labour force in the informal sector in selected developing countries, 1980s*

| City | Share (%) |
| --- | --- |
| Lagos, Nigeria | 50 |
| Kumasi, Ghana | 60–70 |
| Nairobi, Kenya | 44 |
| Calcutta, India | 40–50 |
| Jakarta, Indonesia | 45 |
| Colombo, Sri Lanka | 19 |
| Córdoba, Argentina | 38 |
| São Paulo, Brazil | 43 |
| Bogota, Colombia | 43 |
| Quito, Ecuador | 48 |
| Kingston, Jamaica | 33 |

Source; Todaro (1997).

or below subsistence levels in the latter are much larger and are likely to grow under predicted rates of urbanisation. Income inequalities are often more entrenched and apparent in cities than in the countryside (WRI, 1996) and the cost of living in urban centres is also higher. Generally, urban dwellers have to purchase many items which can be accessed freely or more cheaply in rural areas, such as fuel and building materials. Goods and services are more commercialised in urban centres, and urban residents are more reliant on cash income to secure these, which also brings a certain vulnerability to price rises and falls in income. In rural areas, in contrast, there are more opportunities for subsistence production and foraging, for example. Clearly, as discussed in Chapter 4, these opportunities do not fall equally in all places or to all groups in society. Similarly, the real extent of urban poverty is unlikely to be captured by indicators based on income alone, since factors including assets, the size of household and intra-household relations are also determinants of well-being (UNCHS, 1996).

Because of their poverty, many residents of cities in the developing world live in locations and settlements which are hazardous and detrimental to their own well-being. In addition, as noted in Chapter 2, as poverty retreats into certain locations, often those of high ecological vulnerability, the urban poor may degrade these environments further in the course of securing their basic needs. Fundamentally, the poor are unable to afford the more desirable locations in terms of the inherent or acquired characteristics of the land. Wherever the urban poor are concentrated in cities of the developing world, it is commonly at high densities in areas of low rent.

> Poor groups do not live here in ignorance of the dangers; they choose such sites because they meet more immediate and pressing needs. Such sites are often the only places where they can build their own house or rent accommodation. The sites are cheap because they are dangerous.
> (Hardoy and Satterthwaite, 1989: 159)

**Plate 5.2** *Low-income housing*
- a. **Bangkok squatter settlement**
- b. **Public housing, Harare**
- c. **Tenement blocks, Calcutta**

(a)

Source: David W. Smith, Liverpool University.

(b)

Source: author.

(c)

Source: author.

Regularly such locations are close to hazardous installations, such as chemical factories, and suffer continuous air and water pollution as well as the prospect of sudden fire or explosion. But critically for the urban poor, these locations are close to jobs. As Gupta (1998) revealed, it was the high concentration of low-income people around the Union Carbide Factory in Bhopal which caused so many to be killed or permanently injured (over 3,000 dead and approximately 100,000 seriously injured). In Rio de Janeiro, two-thirds of slum dwellers live on the steep slopes which surround the city. In 1988, excessive rainfall led to flooding and landslides which killed an estimated 289 people and left over 18,000 people homeless (Bartone *et al.*, 1994: 10). Box O highlights the relationship of poverty and vulnerability to hazard and ill-health, in a situation where families live on and around solid waste dumps and children are centrally involved in trying to secure household income through the collection and sale of recyclable items.

The demand for housing in the cities of the developing world far outweighs the supply of formal housing units. During the 1980s, for example, it is estimated that just one formal housing unit was added to the total urban housing stock for every nine new households in the developing world (Hardoy and Satterthwaite, 1989). As a result, the majority of the urban population in such cities is housed in unauthorised informal settlements (variously termed shanty-towns, favelas, bustees, etc.), and in slums or tenements ('legal' developments which become decayed and degraded through overcrowding and poor upkeep). Hardoy *et al.* (1992a) estimate that between 30 and 60 per cent of the population of cities of the developing world live in houses and neighbourhoods which have developed illegally and that up to 95 per cent of all new housing may be being built in this way. Figure 5.7 highlights the range of low-income options for housing and some of the key features of the various forms. In addition to the millions of people for whom accommodation is very insecure or temporary, there are also those who have no home at all, but live on the streets. Although data for cities of the developing world are sparse, it is known that there are over 250,000 pavement dwellers in Bombay, for

## Box 0

---

### *Waste picking in Bangalore*

It is estimated that only between 25 and 55 per cent of all waste generated in cities of the developing world is collected by municipal authorities (UNDP, 1985). Particularly in Asia, much more waste is dealt with by vast networks of waste pickers. In Bangalore, for example, street and dump pickers collect an estimated 500 tonnes of waste daily, compared with just 37 tonnes gathered by municipal workers (WRI, 1996: 112). Many businesses, especially in steel, paper and glass production, depend on recycled raw materials supplied in this way.

However, it is poverty which drives the majority of waste collection in the cities of the developing world; gathering waste from the streets and dumps provides for many the only access to the resources they need for clothing, housing, fuel and work. Often it is women and children who work long hours in unhealthy conditions for the meagre returns. There are many direct environmental hazards associated with waste picking: the materials may include hospital and toxic industrial waste, may be contaminated with faeces, and also may contain sharp objects (Hunt, 1996). The build-up of methane gases also regularly causes fire and the risk of smoke inhalation and burns for waste pickers. Indirectly, waste pickers are exposed to further occupational hazards including harassment and social ostracisation.

Children may face greater health risks through factors including their relative lack of judgement and experience in collection, their inherent vulnerability to air-borne pollution hazard, their potential greater number of years in the activity (starting such work at an early age) and through the long-term impacts on their personality development (Hunt, 1996). In an intensive study of 100 child waste pickers in Bangalore in India, the health status of those children involved in this activity was found to be significantly worse than that of children who were not, as shown in Figure 5.6. The greater worm infestation is likely to be due to the contact with faecal materials. Respiratory tract infections are commonly related to the home environment (poor ventilation and close living), but may be worsened by physical work and also infection from contact with wastes. Lymph node enlargement is usually caused by minor infections, but signs of tuberculosis were also found amongst the waste picking children.

**Figure 5.6** *The health status of waste pickers in Bangalore in relation to non-pickers*

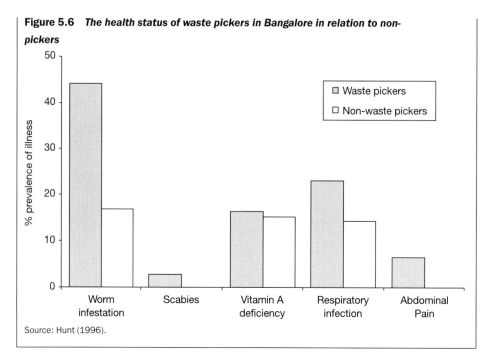

Source: Hunt (1996).

example, and it is likely that the problem is growing generally with the increasing commercialisation of land markets in cities of the developing world.

There is a range of serious disadvantages to living in illegal settlements. For example, in the 1960s and 1970s, many city authorities had policies of slum or squatter clearance, where inhabitants had no defence in the law against eviction (Hardoy and Satterthwaite, 1989). Every year, it is estimated that several million people are evicted from their homes as a result of public works or government redevelopment programmes (UNCHS, 1996: xxviii). Furthermore, because settlements are illegal, they lack basic or emergency public services. Residents may also be ineligible for loans to improve their housing or employment conditions (their illegal shelter or land site being unacceptable as collateral), or indeed, for government subsidies such as in education (for which an authorised address may be required to register children).

## The urban environmental challenge

The level of economic development has been seen to be one factor influencing the nature of environmental problems facing particular cities. With higher levels of economic development, industrial and

**Figure 5.7   *The different kinds of rental and 'owner-occupation' housing for low-income groups in cities of the developing world***

| Type | Characteristics |
| --- | --- |
| Rented room in subdivided inner-city tenement building | Usually very overcrowded and in poor state of repair |
| Rented room in custom-built tenement building | Usually host many more families than built for. Often poorly maintained |
| Rented room, bed or bed hours in boarding house, cheap hotel or pension | Tend to be poorly maintained with a lack of facilities |
| Rented room or bed in illegal settlement | Share problems associated with illegality. Extra problems of insecurity of tenure |
| Rented land plot on which a shack is built | Highly insecure tenure; owner may require them to move at short notice |
| Rented room in house or flat in lower- or middle-income area of the city | Quality may be relatively good. May be located far from jobs |
| Employer-housing for low-paid workers | Quality often poor. Regularly rules against families living there |
| Public housing unit | Many small in relation to numbers living there. Inadequate maintenance |
| Renting space to sleep at work | Usually total lack of facilities for washing/cooking. Lack of security |
| Renting a space to sleep in public buildings | Total lack of security/facilities. Payments to protection gangs or local officials |
| Building a house or shack in squatter settlement | Insecure tenure. Lack of public services. Dangerous locations etc. |
| Building a house or shack in an illegal subdivision | Sites are purchased and have degree of security. Some infrastructure. Often expensive |
| Building a house or shack on a legal land subdivision on the city periphery | Affordable plots on legal subdivisions often far from jobs |
| Invading empty houses or apartments or public buildings | Occupation illegal. Usually no services |
| Building a house or shack in government site and service or core housing scheme | Often far from jobs. Restrictions on employment activities from home. Eligibility criteria |
| Building a shack or house in a temporary camp | Often government's response to disaster situation. Infrastructure and services inadequate |

Source: extracted from UNCHS (1996), but based originally on Hardoy and Satterthwaite (1989).

energy-related pollution become more problematic, as does the inability to deal with wastes (including toxic). Figure 5.8 confirms the pattern of change in a number of environmental indicators with rising economic development.

Clearly, the nature of environmental problems in particular cities will also be influenced by the rate and scale of urbanisation itself and the degree of concentration of such growth. Fast-growing cities may provide particular challenges for planning and management (see, for example, Devas and Rakodi, 1993). However, it should also be noted

**Figure 5.8** *Environmental indicators at different country income levels*

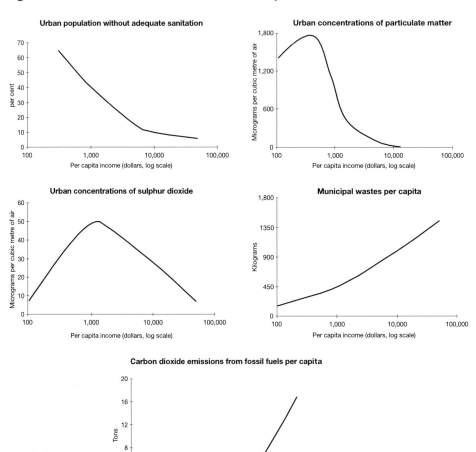

Source: World Bank 1992

that serious environmental problems can also occur in declining industrial centres and stagnant smaller towns, for example (UNCHS, 1996). The geographical location of cities is a further factor shaping the nature of the environmental challenge: cities in cold climates consume greater levels of fossil fuels for domestic heating, for example. Mexico City is a widely cited case where altitude and topology (the city being surrounded on three sides by mountains) combine to present particular challenges for the dispersal of atmospheric pollutants, especially from industry and the motor car. In 1991, for example, Mexico City had only 15 days in the year on which ozone levels were 'satisfactory' according to WHO criteria (World Health Organisation, 1992). Coastal ecosystems (where some of the highest rates of urban growth are currently occurring) also have particular characteristics which are distinct from those further inland, with implications for the nature of environmental problems in cities located on such sites.

However, in order to fully understand the nature of the challenges and opportunities for sustainable urban development, it is necessary to look beyond these broad patterns, to how the various contributing factors interact in specific locations. To this end, substantial insight can be gained through a consideration of environment and development concerns at different levels.

## The household and community level

In poor neighbourhoods of cities in the developing world, many of the most threatening environmental problems are found close to home. Regularly, poorer households use their homes as centres for income generation, their homes also functioning as workshops, as stores for goods for sale, as a shop or as a bar or café. The environmental risks are often greater for women and children because of the longer hours spent at home and in the immediate vicinity. Women, for example, may combine in the same space and time, piece-work for income with domestic duties such as child care. The environmental problems related to such activities in the home are diverse, but include the hazards to health associated with poor ventilation, inadequate light, the use of toxic or flammable chemicals and the lack of protective clothing. A high proportion of disablement and serious injury in cities of the developing world is caused by household accidents and these are strongly aligned to poor quality, overcrowded conditions. For example, many low-income urban dwellings are constructed of flammable materials and accidental fires are more common where families often live in one room and where it is difficult to provide protection from open fires or kerosene heaters (UNCHS, 1996).

Indoor air pollution is also aggravated by the burning of low-quality fuels such as charcoal for domestic heating and lighting. The major impacts are on respiratory health, whereby irritant fumes cause respiratory tract inflammation, repeated exposure leading to the onset of chronic obstructive lung disease. Young children, who may be strapped to their mothers' backs during the course of cooking, suffer more as their smaller lungs are less able to cope with pollutants. In combination with malnutrition, smoke inhalation may further retard infant growth and raise susceptibility of children to other infections (World Health Organisation, 1992).

The most critical determinant of human health (wherever people live) is access to adequate supplies of clean water. In 1980, the WHO estimated that 80 per cent of all sickness and disease world-wide was related to inadequate water (in terms of quantity and quality) and sanitation (services to collect and dispose of solid and liquid wastes). Since then, where environmental improvements have been made in the quality of available water and in the disposal of excreta, illness and the burden of disease have been dramatically reduced and the impact on mortality has been even greater (World Bank, 1992). However, despite the provision of quality water supplies to a further 1.6 billion people during the International Drinking Water Supply and Sanitation Decade of 1980–90, at least 170 million people in urban areas of the developing world still lacked access to potable sources of water in the early 1990s (ibid.: 46).

Table 5.4 presents a regional breakdown of the provision of water and sanitation services in cities of the developing world. Clearly, this scale of analysis masks substantial variation and also subsumes many problems of data quality. Many governments, for example, classify adequate water provision as a tap within 100 metres of the house (WRI, 1996). Such provision may not be enough to ensure sufficient quantities of water at the household level or to secure the health benefits within a community. These may depend, for example, on the numbers of households sharing each tap and the length of time water is supplied each day. To illustrate, in Lucknow, a city in India of 2 million people, water is only available for 10 hours each day (ibid.: 20). In the smaller city of Rajkot, 600,000 people have access to piped water for only 20 minutes each day. When water is restricted in this

**Table 5.4** *Urban water and sanitation coverage in 1994*

| Service | Africa | Asia and Pacific | Middle East | Latin America |
|---|---|---|---|---|
| *Water* | | | | |
| Percentage of population covered | 68.9 | 80.9 | 71.8 | 91.4 |
| Percentage served by house connection | 65.0 | 48.4 | 89.7 | 92.0 |
| Percentage served by public standpipe | 26.0 | 24.0 | 9.3 | 3.3 |
| | | | | |
| *Sanitation* | | | | |
| Percentage of population covered | 53.2 | 69.8 | 60.5 | 79.8 |
| Percentage served by house connection to sewer/septic system | 53.0 | 42.7 | 100.0 | 91.2 |
| Percentage served by simple pit latrine | 22.4 | 8.5 | 0.0 | 5.4 |

Source: WRI (1996).

way, the time taken in queuing and transporting water back to homes serves to limit the amount of water used. For many low-income urban residents, the option is either to draw water from surface sources (often, in effect, open sewers) or to purchase water (of unknown quality) from vendors. In 1989, it was estimated that, for the urban population of the developing world as a whole, 20 to 30 per cent depended on water vendors for their supply (Hardoy and Satterthwaite, 1989). The costs of such water may be anything from four to one hundred times higher than the cost of water from a piped supply (WRI, 1996: 20). This further limits the amounts of water used at the household level and is therefore an important determinant of environmental health.

The inadequacy of urban water supplies generally serves to explain the endemic nature of many debilitating and preventable diseases in cities of the developing world. Diarrhoeal diseases alone killed more than 3 million children in 1993 (WRI, 1996: 21). Vulnerability to infection amongst low-income households is also enhanced by the inadequacies of urban facilities for the hygienic disposal of excreta or household garbage (as summarised in Table 5.4). An estimated 420 million urban residents have no access even to the simplest latrine and therefore have to resort to defecating on open land or in waterways. Such poor sanitation leads to dangers of direct exposure to faeces near homes and the contamination of drinking water. Furthermore, where waste is collected, in 90 per cent of cases it is then discharged untreated and directly into rivers, lakes and coastal waters (Bartone *et al.*, 1994).

The cramped housing conditions of many informal settlements also aggravate the rapid transmission of disease between individuals.

Regularly, as much as 50 per cent of solid waste in neighbourhoods may go uncollected (Hardoy and Satterthwaite, 1989). The subsequent accumulation of domestic refuse and effluent is a further immediate environmental problem in low-income communities. Whilst some waste may be recycled informally, much is left on road sides, dumped in canals and rivers, or left on open lands. Such uncontrolled garbage attracts pests and disease vectors, may present a fire hazard, often contributes to flooding and has direct health risks for children playing in the surrounds of their homes, for example. In 1993, 39 people were killed when a solid waste dump site collapsed in Istanbul and engulfed their homes (Bartone et al., 1994).

In summary, the analysis of urban environmental problems at this level has revealed a number of important and substantial challenges of sustainable development in future. Primarily, it has given a sense of the extent of the environmental challenge in terms of the numbers of people still to be reached by essential improvements if their basic needs of health and livelihoods are to be met and environments conserved. Such understanding points further to the huge economic costs of implementing change, as well as to how sustainable urban development is a long-term and ongoing challenge. This level of analysis also provides an understanding of the reality of urban living, the detail of the Brown Agenda, which is essential for effecting change as considered in later sections.

## The city and the wider region

Although for the majority of urban residents of poor cities, it is the environmental issues closest to home which directly influence their well-being, there are a number of city-wide and regional environmental problems which are mounting, particularly in the rapidly industrialising cities of the developing world, and present further challenges and opportunities for sustainability.

Cities themselves involve the dramatic conversion of land use. Although on a global scale, only around 1 per cent of the total land surface is under urban use (WRI, 1996), urban developments world-wide are encroaching on some of the last remaining and most-valued reserves of natural vegetation, including mangrove swamps, protected wetlands, prime agricultural lands and forests, as noted in Chapter 2.

Urban sprawl is also impacting on existing human activities, such as in eastern Calcutta, where 4,000 hectares of inland lagoons have been filled to provide homes for middle-class families at the expense of the tenant families who formerly made a livelihood based on aquaculture in the region (ibid.). Environmental degradation in peri-urban areas is also occurring through the expansion of unplanned, squatter settlements into areas susceptible to flooding or landslides, for example.

Industrial developments tend to be concentrated in a small number of urban centres as identified in the opening sections of this chapter. Regularly, in the developing world, such developments are not subject to effective planning or pollution control. Indeed, it was seen in Chapter 3 how the lack of stringent pollution control legislation generally in the developing world had been an important factor in attracting industrial production facilities (often multi-national enterprises) to cities in these regions. The types of environmental problems which flow from the location of industry are illustrated in the case of the city of Cubatão, close to Rio de Janeiro in Brazil. This area has been named the 'valley of death' because of its large concentration of Brazilian industrial companies and multi-national firms and the very high resultant levels of air pollution. In Cubatão, levels of respiratory infection, infant mortality rates and the number of still-born and deformed babies are all substantially above those of surrounding regions (Hardoy *et al.* 1992a). Water sources and vegetation in and around the city have also been affected; toxic wastes from these factories have contaminated the major river to the extent that fish are no longer found in it and vegetation has deteriorated from the effects of acid rain. As a result, soils become unstable and residents become vulnerable to further hazards including landslips, often leading to serious loss of life. Box P illustrates the environmental challenge of similar industrial developments at the Mexican border with the United States.

Problems of air pollution have long been associated with cities, although there is currently much diversity world-wide in the relative importance of particular pollutants. Pollution levels can also vary substantially by season. In the developing world, sulphur dioxide pollution and the concentration of suspended particulates are the major causes of urban air pollution, resulting in the main from industrial production and the burning of coal, oil and biomass fuels. In most cities of the more developed regions, tighter environmental regulations, measures to promote more efficient fuel use and the greater use of the least polluting fuels (such as natural gas for

## Box P

### Maquila *developments on the Mexico/United States border*

In the mid-1960s, the government of Mexico initiated a programme to promote industrialisation in the previously underdeveloped northern border region. Mexican and foreign factories were enabled to import machinery and inputs without paying tariffs on the condition that goods were re-exported. US companies were able to take advantage of cheap Mexican labour as well as US tariff regulations. The expansion of export-only factories ('maquiladoras') has substantially altered the distribution of population and urban development in Mexico. The population of the 36 municipalities that adjoin the United States increased from 0.28 million in 1930 to 4 million in 1990 (UNCHS, 1996: 50). Since 1989, maquila industries have also been allowed to sell products in the domestic markets, which coupled with the devaluation of the Mexican currency against the US dollar, led to further urban growth in cities such as Ciudad Juarez and Tijuana (as well as in their 'partner' cities across the border of El Paso and San Diego respectively). The maquiladora zone is also likely to enlarge with the creation of the North American Free Trade deal between the US, Canada and Mexico.

By 1996, there were over 2,000 industrial plants employing more than six hundred thousand people (Dicken, 1998). However, the expansion of employment has often been at substantial cost to the local environment: 'the fact is that Mexican border towns have become garbage dumps for millions of barrels of benzene solvents, pesticides, raw sewage and battery acid spewed out by foreign-owned maquiladoras' (Johnston-Hernandez, 1993: 10). The health impacts of such inadequate disposal of toxic wastes and chemical sludge are profound in urban developments where large proportions of the population depend on open water courses for their drinking water, for example. In the eastern border town of Matamoros, the rate of anencephaly (babies born without brains) is four times the national average, with tissues taken from the mothers of such babies showing the presence of pesticides, several of which have been banned in the United States (Johnston-Hernandez, 1993).

domestic and industrial use) have reduced pollution from the 'traditional sources'. However, city-wide environmental problems also stem from activities other than industrial production. Congested roads and poorly maintained vehicles, for example, are a growing source of 'photochemical' (particularly lead and carbon monoxide) pollution in the developing world as motor vehicle use per capita rises. It is estimated that up to 95 per cent of air-borne lead pollution in cities of the developing world is due to leaded petrol combustion. In many cities, lead levels are in excess of 1.5 micrograms per cubic metre, in contrast to less than 0.8 in most European cities (WRI, 1996: 23) where measures have been taken to promote the use of less polluting (unleaded) fuels, as discussed in Chapter 3.

**Plate 5.3** *Vehicular pollution, Calcutta*

Source: author.

In order to support the resident populations and productive activities therein, cities depend substantially on inputs of raw materials and goods of various natures from the surrounding region. Increasingly, food, fuel and material goods are drawn into cities from all over the nation and indeed the world (Potter *et al.*, 1999). As already suggested, the larger and more prosperous cities make greater demands, as consumption per head rises. The recent idea of the 'ecological footprint' of cities illustrates the demands within cities for renewable resources drawn from outside their boundaries (Rees, 1992) and offers an indicator for comparing the 'sustainability' of different cities. The ecological footprint is a measure of how great an ecologically productive area is required to support city-based production, consumption and waste generation. The methodologies for assessing ecological footprints are under continual modification, particularly as new sources of data become available concerning both the demand on resources and the biological productivity of an area. Generally, core measures to date have tracked the production, import, export and consumption of energy, biotic resources (including food, timber and other crops) and water use, in meeting city needs, including for transport, housing, industrial production, household consumption and waste disposal.

The regional environmental impact of a city is felt in the surrounding area in terms of both the resources on which it draws and the effects of the waste and pollution which it generates. Furthermore, cities have many indirect impacts on rural regions through, for example, the commercialisation of land and agricultural markets in the surrounding areas. This can lead to changes in the type of crops grown and in the nature of productive activities and even to the expulsion of peasant farmers from their lands. Rural to urban migration can have positive impacts on rural economies through wage remittances, for example, but it also often has detrimental consequences in terms of the supply of labour at key points in the agricultural calendar and the loss of entrepreneurial skills, as seen in Chapter 4. Such processes are often linked to environmental decline in the surrounding rural areas.

Clearly, these types of processes illustrate some of the limitations of the dualist distinction of 'urban' and 'rural'. This is confirmed, for example, in the case of urban demand for fuelwood resources, where the supply necessarily comes from forested surrounding areas and, in many instances, pre-empts their use by rural residents (see Soussan, 1990). Sources once available to rural inhabitants become unavailable to them as urban demand rises. This occurs through either deforestation *per se* or the commercialisation of fuelwood, which makes wood a commodity to be paid for rather than a resource to be collected from communal lands. Such regional environmental effects may be felt at considerable distance from the centre of demand (the city). For example, research has shown that fuelwood for the urban population of Bangalore comes from as far away as 140 km. In Delhi, fuelwood comes by rail from Madhya Pradesh (700 km away) at a rate of 612 tons per day (Hardoy and Satterthwaite, 1989). Electrification brings further environmental impacts to rural areas, where large hydro-electric dams, for example, lead to the loss of agricultural area and the displacement of rural people, with the benefits falling largely to urban consumers.

Regional environmental problems associated with city-based activities are regularly linked to the inadequate provision for the safe disposal and dispersal of industrial and domestic waste. In consequence, water, for example, is often returned to sources at far lower qualities than when supplied. Regional impacts include the contamination of domestic water sources and the decline of fishing stocks. Hardoy and Satterthwaite (1989) cite research into water quality in the Bogota river which had faecal bacteria counts so high that it was totally unsuitable for cooking or drinking purposes, even 120 km downstream from the city of Bogota. Similarly, the river passing through Bolivia's largest city, La Paz, has become so polluted that horticultural production downstream has had to be curtailed.

**Plate 5.4** *Child waste pickers*

Source: Hamish Main, Staffordshire University.

In summary, the urban environmental challenge in the developing world is substantial in terms of extent and scope. Industrial developments and rising consumption in urban centres have been seen to be important factors in the degradation of urban environments, but it is also evident that for many low-income households, poverty and a lack of development closely define their core environmental challenges of daily living and working (i.e. the nature of the Brown Agenda). However, it has also been seen that some urban environmental problems diminish as cities become more productive and economically advanced, suggesting that there are also opportunities that cities offer for more sustained development.

Urbanisation, for example, is associated with changed values and lifestyles, including those resulting in smaller families and raised standards of living. Whilst the density of population in urban centres concentrates demand on services and resources, it also offers more opportunities for cheaper delivery of those services and for recycling and waste minimisation efforts than is the case with more dispersed populations. In part as a result of the density of urban living, cities are also places where a great variety of local initiatives and actions develop, outside the formal or monetarised sectors, within citizens' groups, residents' associations, youth clubs, etc., which are increasingly recognised to be essential for 'healthy' cities world-wide and a key resource for sustainable urban development actions.

## Towards sustainable urban development

Figure 5.9 summarises the meaning of sustainable development as applied to cities. The various needs of urban residents and of the environment on which livelihoods depend are highlighted. It is evident that reconciling immediate and future needs as listed is a substantial challenge for action. Indeed, as Bartone *et al.* acknowledge:

Reversing the deterioration of the urban environment without slowing

**Figure 5.9** *The meaning of sustainable development as applied to urban centres*

.................................................................................................................

**Meeting the needs of the present . . .**

\* Economic needs: include access to an adequate livelihood or productive assets; also economic security when unemployed, ill, disabled or otherwise unable to secure a livelihood.

\* Social, cultural and health needs: include a shelter which is healthy, safe, affordable and secure, within a neighbourhood with provision for piped water, sanitation, drainage, transport, health care, education and child development. Also a home, workplace and living environment protected from environmental hazards, including chemical pollution. Also important are needs related to people's choice and control – including homes and neighbourhoods which they value and where their social and cultural priorities are met. Shelters and services must meet the specific needs of children and of adults responsible for most child-rearing (usually women). Achieving this implies a more equitable distribution of income between nations, and in most, within nations.

\* Political needs: includes freedom to participate in national and local politics and in decisions regarding management and development of one's home and neighbourhood – within a broader framework which ensures respect for civil and political rights and the implementation of environmental legislation.

**. . . without compromising the ability of future generations to meet their own needs**

\* Minimising use or waste of non-renewable resources: including minimising the consumption of fossil fuels in housing, commerce, industry and transport plus substituting renewable sources where feasible. Also, minimising waste of scarce mineral resources (reduce use, re-use, recycle, reclaim). There are also cultural, historical and natural assets within cities that are irreplaceable and thus non-renewable – for instance, historic districts and parks and natural landscapes which provide space for play, recreation and access to nature.

\* Sustainable use of renewable resources: cities drawing on freshwater resources at levels which can be sustained; keeping to a sustainable ecological footprint in terms of land area on which producers and consumers in any city draw for agricultural crops, wood products and biomass fuels.

\* Wastes from cities keeping within absorptive capacity of local and global sinks: including renewable sinks (e.g. capacity of river to break down biodegradable wastes) and non-renewable sinks (for persistent chemicals; includes greenhouse gases, stratospheric ozone-depleting chemicals and many pesticides).

.................................................................................................................

Source: UNCHS (1996).

economic development will require an environmental policy strategy that takes into account a wide range of actors, difficult political and economic trade-offs, and a complex set of natural, social, and economic relationships.

(Bartone *et al.*, 1994: 8)

In continuity with the progress shown towards sustainable rural development, if the needs of urban residents world-wide are to be met without compromising the ability of future generations to meet their own needs, change is required throughout the hierarchy of levels of action considered in Chapter 3. For example, one of the most valuable resources available for sustainable urban development is now considered to be the capacity of citizens' groups to 'identify local problems and their causes, to organise and manage community-based initiatives and to monitor the effectiveness of external agencies working in their locality' (UNCHS, 1996: 427). However, the realisation of this capacity depends substantially on what happens at the city authority level, particularly in terms of the establishment of an effective system through which local people (including business interests) can participate in decision-making. In turn, city authorities remain responsible for many functions which are critical to improving urban environments but are widely constrained in the developing world in part through the inadequate transfer of national finances to this level. Yet there are also many issues, such as the reduction of greenhouse gas emissions or promoting more sustainable international trade practices and other essential elements of sustainable urban development, on which city and municipal governments 'cannot be expected to take the lead' (Hardoy et al., 1992a: 191).

## The effectiveness of city authorities

City authorities world-wide are responsible for a range of tasks including regulating building and land use, providing systems of water supply, sanitation and garbage collection, controlling pollution, managing traffic, delivering emergency services, and providing health care and education. Issues of the capacity and responsiveness of local and sectoral institutions are an important determinant of the quality of the environment in a city. Yet many city governments in the developing world are seriously constrained in terms of the finances and professional and technical competencies necessary to provide the investments, services and pollution control central to healthy urban environments. 'City governments in Africa, Asia and Latin America often have one hundredth or (at their most extreme) one thousandth of the revenue per capita available to most city or municipal governments in Europe. Yet, their range of responsibilities is often comparable' (Hardoy et al., 1992a: 162).

In many developing countries, local governments depend on central governments for financial assistance to a much greater degree than in more developed countries. If local authorities are to address their challenges of environmental management more effectively, there is a need to devolve more responsibility for initiating, determining the rate of, and administering systems of local taxation. Until recently, however, most governments of the developing world were centralised, often authoritarian regimes which had sought to consolidate their power through the establishment of (and the concentration of financial resources within) national urban development corporations and national housing authorities, for example. The result was often the construction of large, expensive infrastructural developments in urban centres, but inadequate financial resources at the local authority level to operate and maintain these.

However, effective local government depends on more than increased income. A further legacy of the general failure of central government to transfer management responsibilities to local authorities in the past is the current lack of trained personnel at this level, for example. But perhaps the most fundamental challenge for city authorities in the developing world in the future lies in the way they will work with other organisations at the local level. It was seen in the analysis of the nature of the environmental agenda at the household level, for example, that the extent of the problems and the shortfall in delivering environmental improvements are likely to remain beyond the capacity of local authorities alone to address for many decades. New partnerships are therefore essential to overcome this 'backlog'. Most fundamentally, the analysis also highlighted how the environmental concerns of the poor are intricately linked in the same space and time to economic and social goals. The traditional sectoral policies of urban authorities are ill-equipped to balance such concerns. City authorities are not usually well informed, for example, concerning the extent of the economic, social and environmental advantages of informal activities in waste recycling. Furthermore, understanding of the Brown Agenda has confirmed the capacity for innovation which exists amongst communities when they are given the appropriate support and advice and enabled to participate in democratic systems of governance at the local level.

Establishing new alliances and partnerships and tapping the knowledge and capacities of the local urban population are two core (interrelated) characteristics of 'good city governance' which is now recognised to be a critical condition for sustainable urban development. The failure to develop representative administrative and political systems at the local level is considered to be a primary cause of much environmental degradation to date and good governance is now acknowledged to bring major economic and social gains as well as less environmental degradation in the future (UNCHS, 1996). Over the past ten to fifteen years, progress in this respect has come through wider moves towards national democracy as well as from citizen and community pressure for more effective and accountable city authorities. In many Latin American countries, for example, public election of municipal officers is widespread, but in other regions, particularly on the African continent, the extent of transfer of political power remains very limited (WRI, 1996).

## Utilising the potential of community organisations and local innovation

> Much could be achieved in terms of the direct improvement of living conditions and of services if governments no longer deemed illegal and repressed but instead supported the vast range of individual, house-hold and community-based actions which are the most dynamic force within most cities.
>
> (Hardoy *et al.*, 1992b:152)

There are plenty of examples of communities in urban centres of the developing world over the last decades taking actions to improve their living conditions. Indeed, the total investment by individuals and groups in their homes and neighbourhoods has greatly exceeded that made by city authorities (UNCHS, 1996). 'Often through no choice of their own, low-income households are *de facto* managers of the local environment' (WRI, 1996: 134). But it is only recently that international institutions, aid agencies and national governments have recognised such initiatives as valuable. In the 1950s and 1960s, for example, many national governments (with international backing) engaged in policies of squatter settlement destruction and removal. During this period, self-help housing was viewed with 'alarm and pessimism' (Potter and Lloyd-Evans, 1998: 144) and was seen as part of the problem of underdevelopment (thus necessitating clearance) rather than a reflection of poverty or even part of a solution. Just as with local rural development initiatives which are showing signs of sustainability, understanding is now emerging not only of the value of local initiatives *per se*, but of the preconditions which enable successful urban environmental management based on community organisations to be generated more widely.

**Figure 5.10** *Common characteristics of sustainable urban development*
······················································

1 Housing is also a people's problem
2 The need for building communities
3 The need for organising the community
4 The importance of outsiders
5 The importance of external funding.
······················································

Source: Conroy and Litvinoff (1988).

In the late 1980s, Conroy and Litvinoff presented five core lessons for successful sustainable urban development (as listed in Figure 5.10) on the basis of the work of NGOs and community organisations across twenty human settlement projects (Conroy and Litvinoff, 1988). In continuity with the lessons of sustainable rural development in Chapter 4, they encapsulate often very wide-ranging changes in research and development which have been built on throughout the 1990s.

In short, the first pre-requisite for sustainable urban development is to recognise that housing is not only a problem for central government, local authorities or the private sector but also a concern for communities themselves. '*Given a chance*, poor communities hold the key to the solution of their own problems' (Conroy and Litvinoff, 1988: 252). The nature of that chance and the type of support communities have received have varied. One of the most obvious forms of support is in accessing land, finance and materials for carrying out habitat improvement. Box Q details the experience within a low-income district of Cali (Colombia's second largest city), where an NGO, the Carvajal Foundation, assisted residents by building a warehouse in the middle of the 'squatter settlement' to provide space for manufacturers to sell construction materials directly to residents at wholesale prices (WRI, 1996). Until that time, a major factor in the inability of residents to build and improve houses had been the cost of construction materials which they had had to buy from retailers at some distance from the settlement. Once people had access to the construction materials, they were given further support such as in design and construction. Critically, the Foundation played an important part in convincing the city authorities to approve the building plans and to set up a small office in the neighbourhood: 'Having preapproved building plans and easily obtainable permits was a valuable incentive for residents to build legal, affordable structures' (ibid.: 138). The lesson of experience in Cali (as well as elsewhere) is that securing such support from government authorities is important in giving a sense of stability in a community (essential for encouraging innovation) where the legal right to occupy land may still be lacking.

The second lesson concerning the conditions for sustainable urban development shown in Figure 5.10 encapsulates the way that thinking

## Box Q

___

### *Community innovation in Cali, Colombia*

Cali, a city of 1.7 million inhabitants, is located in a rich agricultural valley and is an important industrial and commercial centre. Many of Cali's residents live in illegal squatter settlements on government or privately owned lands without the required permits. Basic environmental services such as water, sewers, garbage collection, health care and schools are in extremely short supply and poverty is widespread. Aguablanca is one such district, covering 1,500 hectares and home to in excess of 350,000 residents.

The Carvajal Foundation is an organisation which has developed a number of programmes to support small businesses and community initiatives in Aguablanca and other low-income settlements in Cali. The strategy of the Foundation was to observe what people were doing to improve their living conditions and what obstacles they faced in such actions. A major constraint on improvements was found to be the high costs of building a house. The Foundation built a warehouse in the centre of the squatter area from which manufacturers could sell construction materials directly to the residents at lower (wholesale) prices.

With access to affordable materials, many people went on to build houses with assistance from family and friends, but many others lacked knowledge of construction methods. The Foundation enlisted the help of architects to develop a simple, modular house design that was appropriate to the needs of residents and could be developed and extended as finances allowed. The Foundation was keen to involve government agencies in the efforts and lobbied to get the city to approve the standard building plan and to set up an office in the area from which residents could quickly obtain building permits.

Subsequently, the government-owned Central Mortgage Bank also opened an office in Aguablanca, encouraging residents to open savings accounts and offering loans for construction. The Foundation continues to offer training in financial management and in construction, for families interested in building their own homes. The success of the original programme in Aguablanca inspired a private developer to subsequently develop 3,000 plots in another part of the city. In 1992, the Cali city authorities also adopted the same model and launched a programme for 28,000 minimum-wage families. The Carvajal Foundation has since moved also into training for small-scale business, particularly in the food sector.

Source: WRI (1996).

and action have broadened the view of settlement improvements, from the production of physical structures and outputs alone to include the critical issue of social organisation. Enabling communities to get organised is essential for solving the immediate problems and for long-term benefits in the future. Just as with sustainable rural developments, community development itself is something 'more than participation'. It requires working with the poorest and most excluded groups, understanding and addressing their priorities in urban environmental management, as well as bringing together 'different voices' in the community, for example.

Involving women in community development can bring substantial urban improvements. In low-income communities, women regularly take responsibility for providing not only individual consumption needs within the family, but also needs of a more collective nature at the community level, for example, in the provision of basic collective services such as health care, water and education. As such, they regularly engage in community mobilisation activities to ensure that such needs are met. In Guayaquil in Ecuador, low-income settlements are concentrated in the tidal, swampland area of the south and west of the city. Although individual families were able to build houses without major difficulties, accessing basic infrastructural developments required substantial lobbying of municipal authorities over time, activities in which women's efforts were critical (Moser, 1995). For example, women urged their neighbours to form a committee in protest against their appalling living conditions. Although both men and women became members, women took responsibility for much of the day-to-day work of the committee and were the most regular and reliable participants. The women themselves saw it as their responsibility to improve the living conditions of their families through participation in community-level mobilisation. But they also saw themselves as the key beneficiaries of the improvements that were made, in accessing a piped water system or access roads, for example, since many of the domestic tasks such as water collection were undertaken primarily by them (ibid.). However, there were costs for the women; time spent in mobilisation activities often meant compromises had to be made in other domestic and productive activities in order to participate.

> Involving women does not mean that the whole burden of community management should be placed on them. Governments, planners, and even NGOs often make unrealistic assumptions about how much energy, time, and money women can spend in communal or individual self-help programs to improve their environment. Case studies of communal housing projects from Harare, Zimbabwe, to Kingston, Jamaica, to Cordoba, Spain, demonstrate that women who are the sole income earners and childcare providers often do not have the time, skills, or money to invest substantially in community management.
>
> (WRI, 1996: 137)

Enabling communities to became organised in ways which do not burden or exclude particular groups is the third condition for sustainable urban development. Successful projects to date have used various methodologies to encourage social organisation, including women's organisations, but also mobile training teams, established political party frameworks, and new groupings based on a number of houses or streets, for example. All successful projects have shared the common approach of 'learning-by-doing': methods for organisation were flexible and were adapted on the basis of ongoing experience and evaluation of the project. Box R provides detail of the widely reported Orangi project in Karachi, Pakistan, which utilises a number of ways of enabling low-income communities to become organised and take on change. The Orangi case also illustrates many of the other, interrelated elements of sustainable urban development practice. Fundamentally, households do not live in isolation, but look to others in the community for help, in the negotiation of responsibility for collective activities such as neighbourhood cleaning and in determining the allocation of resources. Some of the most successful projects have been those which have built on these existing community networks.

Mobilisation of communities in environmental management has also been found to be successful when linked to income-generation activities. Many recycling projects, for example, have delivered jobs or greater opportunities for sale or exchange of collected items to low-income households with benefits for both the environment and for poverty (Hardoy *et al.*, 1992b). In Cairo, a particular ethnic group, the Zabbaleen, have long earned incomes through waste picking. Over the past decade, several local and international groups have worked to raise these incomes and deliver further environmental benefits through building on the existing community networks. For example, small loans have enabled families to buy machines that convert inorganic wastes such as plastics into useful secondary materials which command higher prices. The construction of local compost plants has also created new jobs in organic waste recycling and simultaneously reduced the amount of garbage on the streets (WRI, 1996).

The fourth pre-requisite of sustainable urban development, the importance of outsiders, has been confirmed in the case studies already referred to. It is evident that enabling communities to improve their own environments depends on various forms of support, social, legal and technical, as much as on access to physical inputs or even finances *per se* (the 'fifth lesson'). Often it has been NGOs which have been central in acting as support, advisory or action groups for community initiatives (or in securing these functions from other

## Box R

### *The Orangi Pilot Project, Karachi, Pakistan*

Orangi is the largest of some 650 low-income squatter settlements in Karachi (WRI, 1996). Karachi itself has expanded from a population of around 400,000 fifty years ago, to a city of over 10 million inhabitants with a further half a million migrants arriving every year:

> most join the 4 million people who live in illegally built squatter settlements. In some areas people live five to a room and 20 per cent of babies do not make it to their first birthday. . . . At the same time, Karachi's political structure is in meltdown. . . . Crime and corruption are rife. . . . Government has palpably failed. . . . The formal sector is defunct.
>
> (Pearce, 1996: 39)

Since 1980, approximately 50 per cent of the estimated 1 million households in this community have been part of the Orangi Pilot Project (OPP), financed largely through the former Bank of Credit and Commerce International's charitable trust and residents' contributions. The project started as a low-cost sanitation programme, but has expanded into other community development efforts including health and family planning for low-income women, housing upgrading, credit schemes for small-scale business and a programme for women's employment (WRI, 1996). In the context of the 'illegality' of the Orangi settlement, and the paucity of state institutions and finances, the community itself had to be the primary source of resources for the development of the most pressing need, that of sanitation provision, and particularly the disposal of excreta and waste water. The primary role for OPP was to provide mainly research and extension services and limited capital inputs. The project in its initial stages focused on finding out what methods for waste disposal were currently being used by the community, selecting the most desirable and building on these to achieve a low-cost method of sanitation within the community.

An underground system of sewerage lines was chosen. Community organisations were created, based on lanes encompassing 20 or 30 houses. Paid employees of OPP held meetings with residents in order to explain the programme and its economic and health benefits. If all of the residents of each 'lane' agreed to participate, they would receive the support of OPP to construct sewerage lines in their area. Each participating unit then elected a 'lane manager' who acted on behalf of the residents in subsequent dealings with OPP. These managers were also responsible for the implementation of the project (including handling of the money collected from the people). OPP was able to reduce the costs of construction through simplifying designs and standardising parts, delivering internal latrines and underground sewerage lines to lanes at approximately one-fifth of the cost of similar improvements made elsewhere by city authorities (WRI, 1996).

OPP staff continue to be involved in undertaking surveys, giving technical advice, organising extension activities, overseeing construction and providing capital expenditure. By 1995, over 94,000 latrines, 5,000 underground lane sewers and 400 secondary drains that carry water to local rivers had been constructed (Pearce, 1996). The people had invested over £1.2 million in this effort and OPP £70,000 (ibid.).

OPP has been successful in providing sanitation to households at a development cost to the user substantially below that which it would have been if such a service had been provided by the Karachi Municipal Corporation. In addition, there are no problems of collecting charges or recovering loans for the local authority and maintenance is undertaken by the residents themselves. Major environmental changes have occurred in the Orangi settlement: human health has improved, people are upgrading their houses and the value of these has gone up. The experience of community participation in a project having clear benefits to individual households has led to improved social harmony in an ethnically very diverse area. The project is rooted in a philosophy of self-help and the capacity of communities to deliver services and promote developments which are likely to be sustained. The lessons of OPP are now being extended into other low-income settlements of Karachi.

Sources: Hasan (1988); Pearce (1996); WRI (1996).

agencies on the community's behalf). As seen in previous sections, it is often the relationship between community and local authorities which is critical for capturing the potential of community organisation and local innovation. NGOs have often enabled communities through adoption of a mediating role with local and central governments (as well as with market forces). In Lusaka in Zambia, for example, a national NGO, Human Settlements of Zambia (HUZA), has acted as a go-between for settlers and public authorities at many stages in a programme of squatter upgrading. The outputs of the programme have included reducing costs of house building, promoting informal activities, and nutrition and health improvements within the community (Conroy and Litvinoff, 1988).

Clearly, such external support requires funding directly in terms of salaries for NGO staff. Funding is also required for the physical and development work that upgrading implies. In the Lusaka case, for example, half of the funds came from a World Bank loan. In the Orangi case in Box R, an international bank covered both the costs of the project teams and the loans to the residents. However, few aid agencies have to date given much attention to the types of urban interventions which are likely to have the biggest impact on the poor (Hardoy *et al.*, 1992a). Where multi-lateral and bi-lateral funds have been forthcoming for urban projects, these have tended to be for large infrastructural projects, for roads, electrification and central hospitals, for example.

Figure 5.11 highlights a number of recent international initiatives on urban environments in the developing world. It is evident that there is a general move on the part of international institutions away from urban projects towards 'institution building' amongst city and regional authorities. The influence of neo-liberal development

**Figure 5.11** *International urban environment programmes*

..................................................................................................

- **The Urban Management Programme (UMP)**

  Funded jointly by the UN Development Programme (UNDP) and a number of bi-lateral agencies, the UMP is administered through the UN Centre for Human Settlements (Habitat), via four regional offices in Accra, Kuala Lumpur, Quito and Cairo. Established in 1990, it aims to assist cities in developing urban environmental strategies and building national and local capacity for management. It also funds research and documents best practice.

- **The Metropolitan Environmental Improvement Programme (MEIP)**

  This programme is funded by UNDP and managed by the World Bank and works specifically towards environmental improvement in Asian cities. It assists cities such as Beijing, Bombay, Colombo and Kathmandu in developing environmental management strategies and action plans. In particular, it seeks to strengthen the institutional and legislative framework for pollution control and environmental protection and to build local environmental networks through workshops, demonstration projects and the dissemination of information.

- **Local Initiative Facility for Urban Environment (LIFE)**

  This programme was launched in 1993 by UNDP in order to directly support local groups in developing countries to improve the urban environment. In particular, collaborative ventures between NGOs, community development organisations, local authorities and the private sector are supported.

- **The Sustainable Cities Programme**

  In 1990, the UN Environment Programme (UNEP) and the UN Centre for Human Settlements initiated a programme which aims to build environmental planning and management capacity at the city, country, regional and global levels. At the city level, demonstration activities are under way in more than 20 countries, mobilising local resources for the development of environmental strategies and implementing priority projects. More broadly, the programme marshals finances from bi-lateral and multi-lateral sources and promotes the sharing of information between participating cities and countries. It is closely aligned to the MEIP and UMP.

- **Healthy Cities Project**

  This World Health Organisation project was started in 1986 and aims to serve as a forum through which all interested parties including city officials and NGOs can exchange ideas and best practice in urban environmental issues, ranging from traffic, through housing to mental health, for example. The first networks were established in Europe, but have since expanded internationally to include hundreds of cities around the world.

..................................................................................................

Sources: WRI (1996); Bartone *et al.* (1994).

thinking (as discussed in Chapter 3) is evident in the priority now given to decreasing the role of the state and improving and building national, regional and local capacity in urban planning and management. New partnerships to deliver services and promote wider development are also now central in these international programmes and publications. However, as seen in the sections above, many cities in developing countries are facing the challenge of sustainable urban development with inadequate capital budgets and in the context of stagnant or deteriorating national economies. The quality of city environments in the future will therefore also depend on changes in the international economic system to promote greater economic prosperity and stability in the countries of the developing world.

## Conclusion

Whilst the Green Agenda has tended to dominate Western environmental thinking and the actions of international institutions of development such as the World Bank and the United Nations, there is now a better understanding that it is the immediate adverse effects on survival for the urban poor of such basic processes as cooking, washing and working which ensure that the environmental challenges at the household level are of no less global proportions than global warming itself.

Sustainable urban management in developing countries evidently requires interdependent actions across all levels of the hierarchy. If the actions of community groups are to be replicated widely to deliver environmental improvements on this scale, decentralised and democratic city and national governance is essential for enabling local groups to organise. External assistance is required at all levels: in building capacity and competence amongst local authority planners and in fostering consensus and leadership within communities. All actions depend on new partnerships built on new approaches to understanding as well as new kinds of interventions.

In continuity with the lessons learned about promoting sustainable rural livelihoods, urban development in the future must focus on the welfare needs of the poorest sectors of the towns and cities of the developing world. The urban environments of the poor are extremely hazardous to human health and the people themselves represent a substantial resource for the improvement of these environments. Enabling poor communities to take control of their own development is the starting point for achieving levels of urban development and environmental change which are less likely to be met by international and/or government finances. Very often this also involves safeguarding the needs of specific groups within poor communities against more powerful economic interest groups.

It is also evident that the Brown Agenda of the majority of urban residents of developing countries encompasses challenging issues of the immediate future which are explicitly interdependent. They are simultaneously about generating an income to live day by day and the reality of resource degradation and danger, for example. Unemployment is closely related to poverty and in turn to hazardous and deteriorating living and working conditions. The challenge of sustainable development includes a shift away from narrow sectoral programmes in urban development towards approaches which can address and utilise these interdependent concerns.

Furthermore, lessons are also being learnt within the developed and developing worlds that a well-functioning urban system also depends on social stability, equity, integration and justice (UNCHS, 1996). Sustainable urban development requires new policies which reduce poverty and other forms of deprivation, but which are also socially inclusive. These are further factors lying behind the essential need for improved local governance to ensure that city authorities address specific local needs and are accountable to all citizens within their jurisdiction.

This chapter has considered the major constraints on and necessary conditions for sustainable urban development at a variety of levels, from community organisation to international political and economic activities. However, prospects for sustainable urban development are also tied closely to those of securing sustainable rural livelihoods. Rural-to-urban migration remains an important force in urban growth. Policy and practice in such areas as economic support for agricultural produce and urban food pricing will therefore have an important effect on the movement of people from rural areas to the towns and cities. The challenges and opportunities of sustainable development lie in providing security for individuals in meeting their basic needs in urban and rural areas; only then will they be able to take a long-term view of development and the environment.

# Discussion questions

* Compare and contrast the key environmental challenges of cities of the developing world with those in more developed regions.
* Examine the ways in which the 'pollution of poverty' is illustrated within the Brown Agenda of low-income households in cities of the developing world.

* Discuss as fully as possible why city authorities are critical in determining the prospects for sustainable development.
* Research evidence for the suggestion that communities themselves can and should play a key role in sustainable housing development.

# Further reading

Conroy, C. and Litvinoff, C. (1988) *The Greening of Aid: Sustainable Livelihoods in Practice*, Earthscan, London.

Hardoy, J., Mitlin, D. and Satterthwaite, D. (1992a) *Environmental Problems in Third World Cities*, Earthscan, London.

Potter, R.B. and Lloyd-Evans, S. (1998) *The City in the Developing World*, Addison Wesley Longman, Harlow.

UNCHS (United Nations Centre for Human Settlements) (HABITAT) (1996) *An Urbanising World: Global Report on Human Settlements*, Oxford University Press, Oxford.

WRI (World Resources Institute) (1996) *World Resources, 1996–97: The Urban Environment*, Oxford University Press, Oxford.

**Sustainable development in the developing world: an assessment**

## Introduction

The preceding chapters of this book have presented a number of insights into the challenges, opportunities and progress made with regard to sustainable development in practice in the developing world. The idea of sustainable development has been seen to be a strong influence in directing wide-ranging changes in development interventions: in terms of the way individuals act, business interests operate and communities organise themselves, for example, but also in directing the nature of state activities, in prompting the formation of new international institutions, and in fostering new ways in which all such organisations relate to each other in the search for patterns and processes of change which are sustainable. Whilst the progress made towards sustainable development is substantial, the challenges have also been seen to be ongoing and indeed are evolving continuously. They also remain contested.

## A common future?

The global nature of sustainable development was identified in Chapter 1. In short, unprecedented rates and degrees of environmental, economic and political change currently impact on and connect people and regions across the globe. The detail of these processes of change and interaction (which constitute the challenges of sustainable development of Chapter 2) confirmed that suffering from environmental degradation is almost universal in the sense that contemporary pollution has an international impact and local pollution problems are repeated the world over (Yearley, 1995). Furthermore, deprivation in its many guises, particularly of poverty and social exclusion, is also a problem which affects individuals, communities and countries world-wide.

Whilst in subsequent chapters the priority was to focus predominantly on the particular environment and development challenges and progress in the developing world, the interdependence of different regions of the world was demonstrated repeatedly. Fundamentally, change in any one element of a global economic system has implications for the functioning of the system as a whole. So it was seen in the rural sector that changes in the wider global economy affected agricultural production in nations and communities of the developing world. Problems in agriculture across these scales also affected other economic sectors.

However, it was also evident that there remains substantial dispute in key arenas, particularly at the international level, regarding the appropriate actions to be taken towards sustainable development. During the UNCED proceedings, for example, the protracted negotiations between governments over the climate convention reflected what Redclift (1996) suggests are the 'mutually incompatible' demands of the industrialised and developing regions. Whilst the Northern countries focused on the conservation of the 'sinks' (i.e. the tropical rainforests, largely), those countries of the South wanted the causes of climate change to be tackled. At the other end of the spectrum, there were many examples given in the text of community groups contesting their priorities in development, such as rights to land, against other (usually more powerful) interests. Certainly, there were often clear differences seen in the chapters between the environmental concerns of the poor and those of wealthier citizens, particularly of the industrialised countries. As Redclift (1992) states,

> In urbanised, industrial societies, relatively few people's livelihoods are threatened by conservation measures. The 'quality of life' consider-ations which play such a large part in dictating the political priorities of developed countries surface precisely because of the success of industrial capitalism in delivering relatively high standards of living for the majority (but by no means all) of the population. In the South, on the other hand, struggles over the environment are usually about basic needs, cultural identity and strategies of survival.

(ibid.: 26)

It has been argued that the success of 'mainstream' sustainable development lies in its compatibility with Northern economic development theory and practice (Adams, 1995). The suggestion is that the solutions or principles for practice as identified by the WCED or at UNCED are reformist rather than radical. They have focused on minor changes to the existing economic system, on better planning and on the more careful use of capital, for example. As such, the argument is that the conception of both development and the environment in the late 1990s remains heavily technocentrist, industrialist and modernist as it was in the 1950s and 1960s. The suggestion is that the 'new' orientation and commitment towards 'sustainable development' has not substantially threatened the *status quo* (i.e. the concentration of power in the North). Certainly, it is possible to remain sceptical concerning progress towards substantially new processes of development which reflect a common future for society. For example, it is estimated that two-thirds of the predicted benefits of the GATT/WTO Uruguay signing are due to go to the industrialised world (Watkins, 1995). It was also detailed in Chapter 3 that there were hopes for the Global Environment Facility as an opportunity to address the reform of development assistance from the developed to the developing world generally, which it seems has not been realised.

However, capitalism at present, with all its strengths and weaknesses, remains the framework within which actions for sustainable development have to be taken. Some may continue to argue that the multi-lateralism identified by the WCED as a condition for addressing a common future will not be achieved within such a system of societal organisation and economic production. However, it is also evident within substantive chapters of this book that the system is sufficiently flexible to enable multiple views of environment and development and diverse solutions in practice. It was seen, for example, that many of the Southern NGOs and people's organisations have been spawned through or have developed to foster actions based on more ecocentric views of society's relationship with nature than identified within the mainstream fora and publications. Such activities are prompting a more widespread reassessment of the criteria for success in the exploitation and protection of the resources of the world, of notions of well-being and the significance of non-economic values in sustainable development.

## Growth and poverty

Reviving economic growth was central to the recommendations of the WCED. However, the quality of such growth had to change (becoming less energy-intensive, for example) and it needed to be more equitable to deliver basic needs to the poorest sectors of society across the globe. Under such circumstances, the WCED was optimistic that simultaneous environmental improvements could be achieved. At the end of the 1990s, economic problems are certainly still to the fore, although there are questions as to whether they have in fact become disconnected once more from environmental issues. For example, commentators such as Middleton *et al.* (1993), Kirkby *et al.* (1995) and Redclift (1996) have pointed to the lack of concern at UNCED about poverty and distributional issues generally in the early 1990s: 'In fact, as a comparison between the report [WCED] and the conference [UNCED] shows, the lack of relationship between them is bewildering, when one is explicitly a follow up to the other' (Kirkby *et al.*, 1995: 11).

The suggestion is that, in an era of world economic recession at the end of the century, the pursuit of economic growth has overtaken concerns for equity and the environment. Certainly, evidence that the liberalisation principles of economic restructuring have changed the nature of growth, served to enhance the resource base or empower the poor is hard to find. Indeed, the contemporary crises within major economies such as those in South-East Asia and Latin America suggest that the economic gains which have been made in the past decade are themselves short term and can be very quickly reversed, as witnessed by the massive outflows of capital which preceded the economic crisis in Indonesia in 1998, for example. But the major 'lesson' of sustainable development identified in the course of the book is that the environment is not a 'luxury' that can be tacked on or taken up only in good economic times. Deepening poverty is one of the main driving forces behind not only environmental degradation, but also civil conflict, the rising numbers of refugees and mounting social tensions, which make economic growth harder to achieve.

On the positive side, the rapidity and degree of economic change in areas of the developing world in the late 1990s have forced a widespread recognition amongst government leaders world-wide that new forms of economic adjustment are needed urgently, tailored to the specific conditions and needs of particular countries. The current reassessment of the prescriptions of the International Monetary Fund constitutes an opportunity to consider also the environmental impacts of these policies which were not anticipated when they were formulated.

> Nobody today questions the case for economic reforms aimed at reducing destabilising budget deficits, establishing realistic currency alignments, and restoring balance-of-payments viability. The challenge is to reach these objectives in a manner *which protects the vulnerable*, is socially inclusive rather than exclusive, and *which establishes a foundation for sustainable economic growth.*
>
> (Watkins, 1995: 10, emphases added)

## Financing sustainable development

The UNCED secretariat estimated that an extra $125 billion every year in aid would be required to implement its action plan (Agenda 21) world-wide. Depending on the method of calculation, this was at least a doubling of financial assistance from the industrialised countries to the developing world (Bramble, 1997). The suggestion was that there are substantial start-up costs associated with moves towards sustainable development, such as in the establishment of improved regulatory and enforcement systems. Financial resources are also needed to cushion the short-term impacts on particular groups and to overcome investor reluctance, for example. In the developing world, it is quite evident that financial resources are limited, and therefore it was considered necessary by world leaders at UNCED to provide donor assistance to help cover these 'additional' costs. In the event, however, there has been a drop in official development assistance (ODA) since the Rio conference. Furthermore, attempts which have been made to impose a 'green conditionality' in aid, i.e. the linking of flows of ODA to projects, programmes and policies which have specific environmental ends, as detailed in Chapter 3, have come under criticism, particularly as regards whose interests are really served by projects such as the creation of biosphere reserves.

However, official aid disbursements are no longer the principal source of investment finances in the developing world. As noted in Chapter 1, private capital flows world-wide, mainly in the form of direct foreign investment, have exceeded ODA many times over in the 1990s. Indeed, it is private foreign investment which has been the driving force behind the increasingly integrated international economy characteristic of the last years of the twentieth century. However, such investment in the developing world has been concentrated in a small number of countries in South-East Asia and Latin America (and is proving to be highly speculative and transitory). Furthermore, trade itself is increasingly taking place within *companies*, the power of which was seen to exceed the total income of the countries in which they make their investments.

If more countries are to share more equitably in global prosperity and trade is to work to create economic opportunities in the developing world and support secure and sustainable livelihoods, further reforms to the international trading system are evidently required. In the short term, for example, for many of the world's poorest countries, higher and more stable prices for commodities would bring substantial benefits for individual producers and national economies. However, existing policies which are supporting agricultural over-production and the dumping of subsidised exports by the industrialised countries, or which foster 'resource-mining' in heavily indebted countries, and encourage further expansion of supply of primary products within the developing world (such as within structural adjustment programmes), work against such a scenario at present.

## Conclusion

Issues of the low levels of economic development, the continued recession in the world economy, the problems of mounting debt and the challenge of finding policies to foster new economic opportunities in ways which enhance human welfare and the natural resource base, are substantial. However, many of the approaches to sustainable development advocated in the chapters of this book do not necessarily require large injections of finance. Rather, many are based on local resources and participatory actions, and are low cost. Indeed, the empowerment of local communities is a necessary condition of sustainable development: too often in the past, development processes have in fact served to undermine local control over the resources on which they depend and led to increased insecurity of livelihood and environmental degradation.

This is not to suggest that the challenges of sustainable development are confined to the local level. Transferring and creating power and control within community groups in developing countries depends on many interrelated actions across the hierarchy of levels as identified in Chapters 4 and 5. Whilst such actions may not be expensive, they do require difficult political decisions and often profound (and unsettling) changes at the individual level. This applies as much to consumers in the more industrialised countries, professionals in development research and amongst government leaders in international and national negotiations, as it does to city planners, NGO staff, and to men and women in the communities of the developing world themselves. The many examples given in the text, of actions which are showing signs of being more sustainable, are surely cause for optimism for the future.

## Discussion questions

*   Consider in more detail the suggestion that the capitalist system and resource conservation are fundamentally contradictory.
*   Undertake some research of your own. In what sense do your friends, family, colleagues or contemporaries consider their future to be in common with that of residents of the developing world?

*   Consider the principal challenges of sustainable development in your immediate local area. What changes would be required to foster greater community action towards sustainable development?
*   Has modern environmentalism in the more industrialised world 'had its day' in the light of the continued global economic recession into the late 1990s?

## Further reading

Adams, W.M. (1995) 'Green development theory? Environmentalisms and sustainable development', in Crush, J. (ed.) *The Power of Development*, Routledge, London, pp. 87–99.

Dodds, F. (ed.) (1997) *The Way Forward: Beyond Agenda 21*, Earthscan, London.

Pepper, D. (1996) *Modern Environmentalism: An Introduction*, Routledge, London.

Redclift, M. (1996) *Wasted: Counting the Costs of Global Consumption*, Earthscan, London.

Watkins, K. (1995) *The Oxfam Poverty Report*, Oxfam, Oxford.

# 7 Conclusion

The achievements made towards sustainable development as detailed in this book have been secured without revolutionary political and economic change in the world system (although many national frameworks in Eastern Europe and the former USSR have undergone fundamental restructuring). These achievements have, however, involved a reassessment on behalf of individuals, nations, NGOs and international institutions of the constraints and opportunities of development and the environment (both natural and structural) and of the criteria for success in the exploitation and protection of the resources of the world.

The shortfalls which have also been identified are confirmation that continued progress depends on factors that include political will and government intervention 'into a dynamic economic system whose overriding motor drive is the desire to maximise profits' (Smith, 1991: 272–3). Ensuring that governments do indeed take the required and appropriate actions for sustainable development demands continued public pressure from individuals, NGOs and representatives of business, commerce and industry. Such pressure depends in turn on individual understanding of the challenges and opportunities of sustainable development and includes a personal reassessment of needs and wants, of commitment to the wider community and of obligations to future generations. Although the power of individuals to respond and take action on the basis of this understanding varies, as has been noted, all are to some degree capable of changing what they do even in small ways (Chambers, 1983).

The WCED report was important in giving direction to governments and individuals concerning the changes in thinking and action required for sustainable development in the future. The Earth Summit conference has similarly been important in confirming commitment at the highest level and in reinvigorating private and public awareness of the nature and extent of the challenges of sustainable development.

Assessing progress towards the achievement of more sustainable development processes is a challenge in itself. For example, whilst increasing the material wealth of those in poverty is one 'quantitative' dimension of sustainable development, economic well-being is now recognised to encompass also 'qualitative', non-economic values such as liberty, dignity and spiritual well-being. The latter dimensions are the criteria for successful sustainable development which have regularly been overlooked within past processes of development and have resulted in the patterns of poverty and environmental deterioration described in this book. Progress towards the more qualitative aspects of sustainable development is hard to ascertain: they are long term and multi-faceted, but nevertheless essential for ensuring the ecological, social and cultural potential for supporting economic activity in the future.

Critically, the quantitative and qualitative dimensions of sustainable development are inseparable and mutually reinforcing. The examples in this book of successful sustainable development projects confirm the positive synergism to be gained from prioritising local knowledge and needs in programmes which enable communities to improve their own welfare and that of the environment. The outcome is seen to be greater than the sum of the parts: benefits are gained from addressing the environment and development together which have not been achieved through separate programmes. For example, within such projects people gain security against becoming poorer and, in so doing, achieve the power to participate in further change. It is these characteristics of success which give optimism for future sustainable development.

However, there can be no single or neatly defined prescription for change. There are no 'blueprints' for sustainable development: sustainable development actions depend on embracing complexity itself. It is not possible, for example, to predict what the likely interests of future generations will be, how much technological progress will be made or the precise outcome of global warming. Flexible solutions are required as the nature of the 'problem' evolves and as policies, programmes and projects proceed. However, these are not justifications for inaction today. The obligation of the current generation is both to exploit and to protect the resources of the world in ways which do not preclude options for such actions tomorrow.

# Glossary

**Absolute poverty**  The condition in which people are able to meet only their bare subsistence essentials (of food, clothing and shelter) to maintain minimum levels of living.

**Agenda 21**  The action plan developed for and accepted at the UNCED conference in 1992, towards achieving sustainable development globally.

**Agribusiness**  The organisation of agro-food production, development, retailing and consumption, and the political and economic relationships that underpin these processes.

**Agroecosystem**  The outcome of human manipulation and transformation of ecosystems for the purposes of the production of food and fibre.

**Agroforestry**  The cultivation of annual crops interspersed with perennial trees and shrubs simultaneously on the same piece of land.

**Basic needs**  The basic goods and services (food, shelter, clothing, education and sanitation, for example) needed to meet a minimum standard of living.

**Biodiversity (biological diversity)**  The variety and variability among living organisms and the ecological complexes in which they occur.

**Biomass energy sources**  The combustion of plant and animal materials and residues for heat and light.

**Biotechnology** The interaction of the fields of biology and engineering. Embraces diverse and complex processes and technologies including 'conventional' techniques of propagation, the cloning of tissue cultures and selective breeding, as well as the manipulation of genetic material, gene transfer and the production of transgenic organisms (the creation of 'custom-made' genes).

**Brundtland Commission** *See World Commission on Environment and Development.*

**Capitalism** An economic system characterised by private ownership in which prices are driven by the forces of supply and demand.

**Carbon sinks** Sources in the atmosphere, oceans and vegetation at which processes of absorption, dispersion and photosynthesis circulate carbon and emit oxygen.

**Chlorofluorocarbons (CFCs)** A class of artificial, relatively inert compounds containing atoms of carbon, fluorine and chlorine, developed largely to provide aerosol propellants and refrigerants. Are major *Greenhouse gases* and contributors to ozone depletion.

**Colonialism** The political and legal domination of a society by another *State*.

**Commercial energy** Fuels such as oil, coal, gas and electricity which have a commercial value and are often traded between countries.

**Commission on Sustainable Development (CSD)** An international forum established in 1992 by the *United Nations* General Assembly to monitor progress towards implementing *Agenda 21* (within and beyond UN organisations).

**Communism** An economic system characterised by collective ownership (such as through governments or community organisations) of the means of production, where the use and allocation of resources are regularly set by centralised planning authorities rather than the price system (as under *capitalism*).

**Convention on Trade in Endangered Species (CITES)** Concluded in Washington in 1973, the convention aims to control and/or eliminate international trade in those species (and products deriving from them) whose survival is threatened by such trade.

**Currency devaluation** A lowering of the official exchange rate of a country's currency in relation to all other currencies.

**Debt burden** The amount of interest and repayments due in relation (usually) to the amount received through exports.

**Debt crisis** The mounting problems faced by developing countries in the 1980s in meeting their debt obligations as a result of factors including rising oil prices, increased interest rates on borrowing and declining terms of trade for goods from the developing world.

**Debt for nature swaps** The selling on of a loan debt by a bank at a discounted price to an intermediary (typically an environmental group) which then arranges for the debtor country to pay it back in local currency through conservation and environmental work to that value.

**Dependency school** A set of ideas about development which emphasises the unequal relationships between the developed and developing world, in which the former is seen to extract value from the latter, leading to ever greater disparities between developed and underdeveloped countries.

**Development** The process by which people's level of living, their quality of life and their capacity to participate in the political, social and economic systems and institutions which influence their dignity and freedom, improve.

**Doubly Green Revolution** The search for ways of further raising agricultural production in the developing world without the negative environmental and social consequences associated with the *Green Revolution* of the 1970s.

**Ecocentric** A way of thinking about humanity's relationship with nature which puts the Earth at the centre and searches for alternative ways of living and appropriate technologies consistent with ecological principles and global stability. *See also Technocentrism.*

**Ecological footprint** A measure of the sustainability of a society or region relating the population to the ecologically productive area required to support all associated activities of that settlement.

**Economic growth** Rising levels of national output and income as a result of the increase in productive capacity of the economy over time.

**Empowerment** A desired process whereby (particularly impoverished and marginalised groups) become the agents of their own development.

**Entitlements** The claims and access which individuals and groups have over resources arising from their ownership and/or particular social relationships, including legally recognised private ownership and socially sanctioned traditional community management.

**Environmental determinism** The doctrine that human activities and characteristics are determined and controlled by the environment.

**Environmental Impact Assessment (EIA)** The critical investigation of the likely environmental, social, economic and cultural effects of development projects (and policies in some cases). Accepted practice includes the consideration of alternative development options and locations and the nature of possible mitigating measures. *See also Strategic Environmental Assessment.*

**Environmentalism** The ideas, language and practices which stem from a concern for human–environment relations.

**Fertility rate** The yearly number of children born alive per 1,000 women within the childbearing age bracket (normally 15 to 49 years).

**Foreign aid** The international transfer of public funds in the form of loans or grants from one government to another (bi-lateral aid) or through international organisations such as the *World Bank* (multi-lateral assistance).

**Foreign Direct Investment (FDI)** Investments made overseas by transnational companies.

**Gender** The socially constructed and culturally ascribed difference between males and females that masks particular expressions of masculinity and femininity.

**General Agreement on Tariffs and Trade (GATT)** Created in 1948 as one of the Bretton Woods institutions designed to structure the post-war recovery of Europe. GATT aimed to promote free trade and prevent protectionism which had plagued the international economy prior to the Second World War. Replaced by the *World Trade Organisation* in 1995.

**Global Environment Facility (GEF)** Initiated in 1991 to create new funds for governments of low-income countries to take on environmental actions with clear global benefits in areas such as greenhouse gas emissions and biodiversity. Managed jointly by *UNEP, UNDP* and the *World Bank.*

**Global warming** The apparent warming of the Earth's surface as the concentration of *Greenhouse Gases* in the atmosphere has risen.

Considered to threaten many of the life-support functions of the Earth including disruption of sea levels (due to thermal expansion of the oceans), the melting of ice sheets and enhanced rainfall.

**Globalisation** Common processes of economic, political, cultural and environmental change which are leading to the increased connectedness of different parts of the world.

**Green Revolution** The export of packages of high yielding seed varieties, fertilisers and pesticides to raise agricultural productivity, most widely from core research stations to Asian countries from the 1970s onwards.

**Greenhouse gases** Gases in the atmosphere which partially absorb long-wave terrestrial radiation emitted by the warm surface of the Earth. The main greenhouse gases are water vapour, carbon dioxide, methane, nitrous oxide and ozone. The rising concentration of carbon dioxide, methane, nitrous oxide and *chlorofluorocarbons* is believed to lead to *global warming*.

**Gross Domestic Product (GDP)** An estimate of the total value of all materials, foodstuffs, goods and services that are produced by a country in a particular year.

**Gross National Product (GNP)** Similar to *GDP* but includes also the value of income residents of the country receive from abroad (less similar payments made to non-residents who contribute to the domestic economy).

**Group of 7/8 (G7/8)** The seven leading industrialised nations (the United States, Canada, Great Britain, France, Germany, Japan and Italy) who meet annually to discuss global economic issues. This includes Russia as the eighth member.

**Hazardous waste** Materials which are damaged or superfluous to use which present a threat to human and environmental health. The hazard may lie in the material's toxicity (i.e. is poisonous), its concentration (*CFCs*, for example, are non-toxic but pose a serious ozone depletion hazard), or the method of handling (including whether protective clothing is available, and contact with other products or processes).

**Heavily Indebted Poor Countries Initiative (HIPC)** An initiative of the *World Bank* and *IMF*, supported by the Commonwealth and G7/8 countries, aims to reduce debt to more 'affordable' levels. Requires proven commitment to *IMF* macro-economic reforms.

**Human Development Index (HDI)** Developed by the *United Nations Development Programme* in 1990 to try to produce a more comprehensive measurement of development which accommodates factors of social progress and improvement as well as economic growth (the latter as more conventionally reported by the *World Bank*, for example).

**Human immunodeficiency virus (HIV)** The virus which causes Aids (Acquired Immune Deficiency Syndrome), which is a major killer in the developing world – particularly in Africa – and has no known vaccination or cure.

**Industrial agriculture** The integrated network of industrial processes characterising modern agricultural production, storage, processing, distribution, marketing and retailing.

**Infant Mortality Rate (IMR)** Deaths of children between birth and aged one year old, expressed per 1,000 live births.

**Informal sector** The part of the urban economy (the term is most widely used in relation to employment and housing) characterised generally by free entry, unregulated markets, small scale, labour intensive, petty trading, family firms and indigenous skills.

**Intercropping** The practice of mixing different crops within the same cultivated area.

**International division of labour** The specialisation of workers from particular countries in particular kinds of economic activities.

**International Monetary Fund (IMF)** A financial institution created in 1944 to regulate the international monetary exchange system. Lends money to Member States to alleviate short-term balance of payment problems.

**International Tropical Timber Organisation (ITTO)** Established in 1985 through the work of the *United Nations*, to regulate trade in tropical timber. The first commodity organisation with a conservation mandate. By 1992, had 47 signatory countries which either produced or consumed tropical timber.

**International Union for Conservation of Nature and Natural Resources (IUCN)** Formed in 1956 as a replacement to the International Union for the Protection of Nature established in 1949. One of the first international environmental organisations and an important player in the development of the modern concept of sustainability. See also *World Conservation Strategy*.

**Internet**  A network (or networks) of computers spread all over the world. Integrated sites facilitate access to a wide range of (uncensored, unregulated) data, viewpoints, services and products.

**Land reform**  The process of changing the distribution and scale of land ownership and/or the conditions under which land is held (tenancy arrangements).

**Livelihood**  Stocks and flows of food, cash and income in kind upon which individuals and households live.

**Local Agenda 21**  The translation of national *Agenda 21* commitments into locally defined programmes of action, particularly by local authorities in conjunction with community groups.

**Malnutrition**  Ill-health arising from inadequate diet. Usually measured in terms of average daily protein consumption

**Modernisation theory**  A theory of development which emphasises the stages a country is assumed to pass through as it develops. Development is equated with growth and modernisation, which is contrasted with stagnation and tradition.

**Multinational Corporation (MNC)**  An international corporation with a headquarters in one country and many branch offices (for research and development as well as production *per se*) located in other countries around the world.

**Neo-liberalism**  A set of ideas about political and economic development associated with free market economics, removal of trade barriers and withdrawal of the state. Particularly associated with the ideas of Western leaders such as Ronald Reagan and Margaret Thatcher (see also *International Monetary Fund* and *World Bank*).

**Neo-Malthusianism**  Views on population–resource relationships with origins in the work of Thomas Malthus (1766–1834) and his belief in the inevitability of human starvation and death as population expands geometrically whilst life-sustaining resources increase arithmetically.

**Newly Industrialising Countries (NICs)**  A small group of countries, particularly in Pacific Asia and parts of Latin America, with relatively high levels of economic expansion which developed linkages into world trade networks in the 1980s. Includes what became known as the 'Asian Tiger' economies of South Korea, Hong Kong, Singapore and Taiwan.

**Non-Governmental Organisations (NGOs)** An umbrella term encompassing a range of diverse organisations at various scales which are fundamentally outside government and generally not-for-profit. They range from small burial societies and community welfare or work groups designed to serve their own members, through to international charities working for other people who need help.

**Organisation for Economic Co-operation and Development (OECD)** An organisation comprising the major industrialised countries of Europe and North America, which has the objective of promoting co-operation and economic growth of member nations.

**Organisation of African Unity** Founded in Addis Ababa in 1963 to work for the political and economic unity of independent Africa.

**Overseas Development Assistance (ODA)** *See Foreign aid.*

**Participatory Rural Appraisal (PRA)** May refer to particular techniques/tools for the assessment/analysis of rural areas or to the complete development planning process. Is built on the principle that emphasis should be given to the priorities of rural people themselves, and their full participation in development projects, including setting objectives, planning, execution and evaluation.

**Pollution of poverty** Encompasses the circular relationships between poverty and the environment to include the environmental concerns of poor people (such as ill-health and access to decent shelter, energy sources or basic services) as well as the detrimental impacts of impoverished people on the resource base in the course of living and working. See also *Brown Agenda.*

**Resource scarcity** A mismatch between the supply and demand for a particular resource which may stem from interrelated factors including those of absolute existence, location in relation to areas of demand, price and the activities of regulatory authorities such as government.

**Resource-poor farming** A type of world agriculture distinguished by its location in some of the more ecologically marginal areas of the globe and the poverty and socio-political insecurity of people undertaking such activity.

**Rio Declaration on Environment and Development** The set of 27 principles agreed at *UNCED* in 1992 underpinning the global partnership for environment and development. Many of the principles concern the role, rights and responsibilities of signatory states.

**Social exclusion** The process by which particular groups are deprived of access to socially provided goods and services, including employment, education, health and welfare and political structures. More recently, the concept has been widened to include issues of deprivation and poverty in developing countries.

**Society** The totality of social relationships, interventions and institutions determined by people across particular places and times.

**Sovereignty** The exercise of State power over the resources and people contained within national boundaries and as codified within international laws.

**Strategic environmental assessment (SEA)** The systematic process of evaluating the environmental effects of a development policy, plan or programme (as distinct from the development project which has a tighter spatial and temporal focus). *See also Environmental Impact Assessment.*

**Structural adjustment programme (SAP)** A programme of policy reform commonly defined by the *World Bank* and the *IMF* with the principal aim of promoting economic recovery and market competitiveness.

**Technocentrism** A way of thinking about humanity's relationship with nature which exalts humanity's desire and capacity to manipulate nature through the application of the techniques and products of scientific investigation in particular.

**The Brown Agenda** The environmental concerns of low-income residents of cities of the developing world (such as inadequate sanitation and solid waste disposal services) which have been under-researched and invested in in comparison with the 'Green Agenda' which dominated international environmental thinking and action into the late 1980s (encompassing concerns such as climate warming and biodiversity).

**The state** An independent political unit with internationally recognised boundaries, and the set of institutions and means for control over that specified territory and society.

**Trade liberalisation** The removal of obstacles to free trade such as quotas, tariffs and exchange controls.

**Trans-national Corporation (TNC)** Organisations with investments and operations in many different countries, with no clear 'home-base' or 'headquarters', but rather factories, offices, subsidiary companies, etc., which are integrated at that international scale.

**Underemployment** A situation in which people are working less than they would like and/or their labour is not being fully employed in the sense that they could be removed from the workplace and there would not be a proportional decrease in productivity.

**United Nations (UN)** A global organisation set up at the end of the Second World War with the fundamental intention of working to maintain international peace and security. In recent decades, the bulk of its activity has focused on economic and social development. Almost every country of the world is a member. It is currently a complex web of institutions undergoing substantial reform.

**United Nations Conference on Environment and Development (UNCED)** A conference held in Rio de Janeiro in 1992 to mobilise and direct international co-operation towards sustainable development. Also known as the Earth Summit.

**United Nations Development Programme (UNDP)** A body of the United Nations formed in 1965 with the major function of promoting development through finance and technical support to projects in the fields of health, agriculture and education, for example.

**United Nations Environment Programme (UNEP)** A body of the United Nations located in Nairobi and formed in 1972 to co-ordinate environmental policy within the UN system and to monitor the conservation commitments and actions of Member States.

**Urban bias** The favouring of the urban sector within development policies (at times, unintentionally and at others, by design).

**Urbanisation** Strictly, the proportion of a population resident in areas defined as urban. Also used to encompass the economic (as well as demographic) growth processes of urban centres.

**World Bank (WB)** Common name for the International Bank for Reconstruction and Development (IBRD), which together with the International Development Association (IDA), the International Finance Corporation (IFC) and the Multilateral Investment Guarantee Agency (MIGA) comprise the 'World Bank Group'.

**World Commission on Environment and Development (WCED)** A limited-life group commissioned in 1984 by (but independent from) the United Nations to identify long-term environmental priorities and strategies for the international community. Reported in 1987 (*Our Common Future*) and commonly known by the surname of its Chair, Gro Harlem Brundtland.

**World Conservation Strategy (WCS)** Prepared by *IUCN* in the late 1970s and launched simultaneously in 40 countries in 1980. It identified the main threats to global species and ecosystems and proposed actions and priorities particularly for governments, *UN* agencies and inter-governmental bodies towards accommodating conservation and development.

**World Trade Organisation (WTO)** Formed in 1995 to replace the *General Agreement on Tariffs and Trade*. Administers the agreed international rules on trade policy globally.

# References

Adams, A. (1993) 'Food insecurity in Mali: exploring the role of the moral economy', *IDS Bulletin*, 24,4, pp. 41–51.

Adams, W.M. (1990) *Green Development*, Routledge, London.

—— (1995) 'Green development theory? Environmentalisms and sustainable development', in Crush, J. (ed.) *The Power of Development*, Routledge, London, pp. 87–99.

—— (1996) 'Sustainable development?', in Johnston, R.J., Taylor, P.J. and Watts, M.J. (eds) *Geographies of Global Change: Remapping the World in the Late Twentieth Century*, Blackwell, Oxford, pp. 354–74.

Allen, J. (1995) 'Crossing borders: footloose multinationals?', in Allen, J. and Hamnett, C. (eds) (1995b) *A Shrinking World? Global Unevenness and Inequality*, Oxford University Press, Oxford, pp. 55–102.

Allen, J. and Hamnett, C. (1995a) 'Uneven worlds', in Allen, J. and Hamnett, C. (eds) (1995b) *A Shrinking World? Global Unevenness and Inequality*, Oxford University Press, Oxford, pp. 233–254.

Allen, J. and Hamnett, C. (eds) (1995b) *A Shrinking World? Global Unevenness and Inequality*, The Shape of the World: Explorations in Human Geography series, no. 2, Oxford University Press, Oxford.

Allen, J. and Massey, D. (1995) *Geographical Worlds*, The Shape of the World: Explorations in Human Geography series, no. 1, Oxford University Press, Oxford.

Allen, T. and Thomas, A. (eds) (1992) *Poverty and Development in the 1990s*, Oxford University Press, Oxford.

Barbier, E.B. (1987) 'The concept of sustainable economic development', *Environmental Conservation*, 14,2.

Barrett, H. and Browne, A. (1995) 'Gender, environment and development in sub-Saharan Africa', in Binns, J.A. (ed.) *People and the Environment in Africa*, Wiley, London, pp. 31–8.

Barrow, C.J. (1995) *Developing the Environment: Problems and Management*, Longman, London.

Bartelmus, P. (1994) *Environment, Growth and Development: The Concepts and Strategies of Sustainable Development*, Routledge, London.

Bartone, C., Bernstein, J., Leitmann, J. and Eigen, J. (1994) *Towards Environmental Strategies for Cities*, Urban Environmental Management Paper, 18, World Bank, Washington.

Bernstein, H., Crow, B. and Johnson, H. (eds) (1992) *Rural Livelihoods: Crises and Responses*, Oxford University Press, Oxford.

Biswas, A.K. (1993) 'Management of international waters', *International Journal of Water Resources Development*, 9,2, pp. 167–89.

Biswas, M.R. and Biswas, A.K. (1985) 'The global environment: past, present and future', *Resources Policy*, 11,1, pp. 25–42.

Black, I. (1998) 'Robin Cook's tour of the global badlands', *The Guardian*, 22.4.98.

Bown, W. (1994) 'Deaths linked to London smog', *New Scientist*, 25.6.94, p. 4.

Braidotti, R., Charkiewicz, E., Hausler, S. and Wieringa, S. (1994) *Women, the Environment and Sustainable Development: Towards a Theoretical Synthesis*, Zed Books, London.

Bramble, B. (1997) 'Financial resources for the transition to sustainable development', in Dodds, F. (ed.) *The Way Forward: Beyond Agenda 21*, Earthscan, London, pp. 190–205.

Brown, L.R. (ed.) (1996) *Vital signs, 1996/1997: The Trends that are Shaping Our Future*, Earthscan, London.

Bryant, R.L. and Bailey, S. (1997) *Third World Political Ecology*, Routledge, London.

Buckley, R. (1994) *World Population: The Biggest Problem of All?*, Understanding Global Issues series, 94/7, European Schoolbooks, Cheltenham.

—— (1995) *The United Nations: Overseeing the New World Order*, Understanding Global Issues series, 6/93, European Schoolbooks, Cheltenham.

—— (ed.) (1996) *Fairer Global Trade: The Challenge for the WTO*, Understanding Global Issues series, 96/6, European Schoolbooks, Cheltenham.

Bush, J. (1998) 'Unforgiving formula that leaves life out of the equation', *Financial Times*, 10.4.98.

Cairncross, F. (1995) *Green Inc.: A Guide to Business and the Environment*, Earthscan, London.

Chambers, R. (1983) *Rural Development: Putting the Last First*, Longman, London.

—— (1988) 'Sustainable rural livelihoods: a key strategy for people, environment and development', in Conroy, C. and Litvinoff, M. (eds) *The Greening of Aid: Sustainable Livelihoods in Practice*, Earthscan, London, pp. 1–17.

—— (1993) *Challenging the Professions: Frontiers for Rural Development*, Intermediate Technology Publications, London.

Chambers, R. (1994) 'Foreword', in Scoones, I. and Thompson, J. (eds) *Beyond Farmer First: Rural People's Knowledge, Agricultural Research and Extension Practice*, Intermediate Technology Publications, London, pp. xiii–xvi.

Chambers, R., Pacey, A. and Thrupp, L.A. (1989) *Farmer First: Farmer Innovation and Agricultural Research*, Intermediate Technology Publications, London.

Chissick, R. (1990) 'The gender trap', *The Guardian*, 4.12.90.

Chuta, E. and Leidholm, C. (1990) 'Rural small scale industry: empirical evidence and policy issues', in Eicher, C.K. and Statz, J.M. (eds) *Agricultural Development in the Third World*, second edition, Johns Hopkins Press, Baltimore, pp. 327–41.

Clayton, K. (1995) 'The threat of global warming', in O'Riordan, T. (ed.) *Environmental Science for Environmental Management*, Longman, London, pp. 110–30.

Colchester, M. (1990) 'The International Tropical Timber Organisation: kill or cure for the rainforests?', *The Ecologist*, 20,5, 166–73.

—— (1994) 'The New Sultans: Asian loggers move in on Guyana's forests', *The Ecologist*, 24,2, pp. 45–52.

Commonwealth Secretariat (1989) *The Langkawi Declaration on the Environment*, Marlborough House, London.

Conroy, C. and Litvinoff, C. (1988) *The Greening of Aid: Sustainable Livelihoods in Practice*, Earthscan, London.

Conway, G. R. (1987) 'The properties of agroecosystems', *Agricultural Systems*, 24, pp. 95–117.

—— (1997) *The Doubly Green Revolution: Food for All in the 21st Century*, Penguin, London.

Corbridge, S. (1987) 'Development and underdevelopment', *Geography Review*, September, pp. 20–2.

Craig, G. and Mayo, M. (eds) (1995) *Community Empowerment: A Reader in Participation and Development*, Zed Books, London.

Dankelman, I. and Davidson, J. (1988) *Women and Environment in the Third World*, Earthscan, London.

Davies, S. (1993) 'Are coping strategies a cop out?', *IDS Bulletin*, 24,4, pp. 60–72.

Department of Environment and Natural Resources (1995) *Philippine Environmental Quality Report*, 1990–95, Manila.

Department for International Development (1997) *Eliminating World Poverty: A Challenge for the 21st Century, Department for International Development*, Government Stationery Office, London.

Devas, N. and Rakodi, C. (eds) (1993) *Managing Fast Growing Cities: New Approaches to Urban Planning and Management in the Developing World*, Longman, Harlow.

Devereux, S. (1993) 'Goats before ploughs: dilemmas of household response sequencing during food shortages', *IDS Bulletin*, 24,4, pp. 52–9.

Dicken, P. (1998) *Global Shift: Transforming the World Economy*, third Edition. Paul Chapman Publishing, London.

Dickenson, J., Gould, B., Clarke, C., Mather, C., Prothero, M., Siddle, D., Smith, C. and Thomas-Hope, E. (1996) *A Geography of the Third World*, second edition, Routledge, London.

Dixon, C. (1990) *Rural Development in the Third World*, Routledge, London.

Dodds, F. (ed.) (1997) *The Way Forward: Beyond Agenda 21*, Earthscan, London.

Drakakis-Smith, D.W. (1987) *The Third World City*, Routledge, London.

Dunn, K. (1994) 'Killing the ripest crop', *CERES*, 26,5, pp. 35–411.

Edge, G. and Tovey, K. (1995) 'Energy: hard choices ahead', in O'Riordan, T. (ed.) *Environmental Science for Environmental Management*, Longman, London, pp. 317–34.

Ehrlich, P.R. (1968) *The Population Bomb*, Ballantine, New York.

Elliott, J.A. (1996) 'Resettlement and the management of environmental degradation in the African farming areas of Zimbabwe', in Eden, M.J. and Parry, J.T. (eds) *Land Degradation in the Tropics: Environmental and Policy Issues*, Pinter, London, pp. 115–25.

Farrington, J. and Lewis, D.J. (eds) (1993) *NGOs and the State in Asia*, Routledge, London.

Finger, M. (1994) 'Environmental NGOs in the UNCED process', in Princen, T. and Finger, M. (eds) *Environmental NGOs in World Politics*, Routledge, London, pp. 186–213.

Foley, G. (1991) *Global Warming: Who Is Taking the Heat?*, Panos Publications, London.

Frank, A.G. (1967) *Capitalism and Underdevelopment in Latin America*, Monthly Review, London.

French, H.F. (1990) *Green Revolutions: Environmental Reconstruction in Eastern Europe and the Soviet Union*, World Watch paper, no. 99, Worldwatch Institute, Washington.

Geheb, K. and Binns, J.A. (1997) 'Fishing farmers or farming fishermen? The quest for household income and nutritional security on the Kenyan shores of Lake Victoria', *African Affairs*, 96, pp. 73–93.

George, S. (1992) *The Debt Boomerang*, Pluto Press, London.

Goldsmith, E., Allen, R., Allaby, M., Davoll, J. and Lawrence, S. (eds) (1972) *Blueprint for Survival*, Penguin, Harmondsworth.

Gourlay, K. (1995) 'A world of waste', *People and the Planet*, 4,1, p. 6.

Gupta, A. (1998) *Ecology and Development in the Third World*, second edition, Routledge, London.

Hamnett, C. (1995) 'Controlling spaces: global cities', in Allen, J. and Hamnett, C. (eds) *A Shrinking World? Global Unevenness and Inequality*, pp. 103–42, Oxford University Press, Oxford.

Hanlon, J. and Agarwal, A. (1977) 'Mass sterilisation at gunpoint' *New Scientist*, 74, 1050, pp. 268–70.

Hardoy, J.E., Mitlin, D. and Satterthwaite, D. (1992a) *Environmental Problems in Third World Cities*, Earthscan, London.

—— (1992b) 'The Future City', in Holmberg, J. (ed.) *Policies for a Small Planet*, Earthscan, London, pp. 124–56.

Hardoy, J.E. and Satterthwaite, D. (1989) *Squatter Citizen: Life in the Urban Third World*, Earthscan, London.

Harriss, B. and Crow, B. (1992) 'Twentieth century free trade reform: food and market deregulation in sub-Saharan Africa and South Asia', in Wuyts, M., Mackintosh, M. and Hewitt, T. (eds) *Development Policy and Public Action*, Oxford University Press, Oxford, pp. 199–230.

Hasan, A. (1988) 'Orangi Pilot Projects', in Conroy, C. and Litvinoff, C. (eds) *The Greening of Aid: Sustainable Livelihoods in Practice*, Earthscan, London.

Hayter, T. (1989) *Exploited Earth: British Aid and the Environment*, Earthscan, London.

Hewitt, K. (1997) *Regions of Risk: A Geographical Introduction to Disasters*, Longman, London.

Hewitt, T. (1992) 'Developing countries: 1945–1990', in Allen, T. and Thomas, A. (eds) *Poverty and Development in the 1990s*, Oxford University Press, Oxford, pp. 221–37.

Hildyard, N. (1994) 'The big brother bank', *Geographical Magazine*, June, pp. 26–8.

Hill, A.G. (1991) 'African demographic regimes: past and present', paper presented at the Conference of the Royal African Society, Cambridge, April 1991.

Hirschmann, A.O. (1958) *The Strategy of Economic Development*, Yale University Press, New Haven, Connecticut.

Holmberg, J. and Sandbrook, R. (1992) 'Sustainable development: what is to be done?', in Holmberg, J. (ed.) *Policies For a Small Planet*, Earthscan, London, pp. 19–38.

Houghton, J.T., Jenkins, G.J., and Ephiraums, J.J. (1997) 'The Intergovernmental Panel on Climate Change Scientific Assessment – Policymakers Summary', in Owen, L. and Unwin, T. (eds) *Environmental Management: Readings and Case Studies*, Blackwell, London, pp. 121–47.

Hunt, C. (1996) 'Child waste pickers in India: the occupation and its health risks', *Environment and Urbanisation*, 8,2, pp. 111–74.

ICLEI (1997) *Local Agenda 21 Survey: A Study of Responses by Local Authorities and their National and International Associations to Agenda 21*, International Council for Local Environmental Initiatives.

IDS Bulletin (1998) *Poverty and Social Exclusion in North and South*, 29,1.

IUCN with UNEP and WWF (1980) *World Conservation Strategy*, IUCN, Gland.

Jackson, C. (1995) 'Environmental reproduction and gender in the Third World', in Morse, S. and Stocking, M. (eds) *People and Environment*, UCL Press, London, pp. 109–30.

Johnson, S. (1995) *The Politics of Population*, Earthscan, London.

Johnston, R.J., Taylor, P.J. and Watts, M.J. (eds) (1995) *Geographies of Global Change*, Blackwell, Oxford.

Johnston, R.J. (1996) *Nature, State and Economy: A Political Economy of the Environment*, second edition, Routledge, London.

Johnston-Hernandez, B. (1993) 'Dirty growth', *New Internationalist*, 246, pp. 10–11.

Kelly, M. and Granwich, S. (1995) 'Global warming and development', in Morse, S. and Stocking, M. (eds) *People and the Environment*, UCL Press, London, pp. 69–107.

Kirkby, J., O'Keefe, P. and Timberlake, L. (eds) (1995) *The Earthscan Reader in Sustainable Development*, Earthscan, London.

Knox, P.L. and Marston, S.A. (1998) *Places and Regions in Global Context*, Prentice-Hall, New Jersey.

Koch, M. and Grubb, M. (1997) 'Agenda 21', in Owen, L. and Unwin, T. (eds) *Environmental Management: Readings and Case Studies*, Blackwell, London, pp. 455–9.

Korten, D.C. (1990) *Getting to the Twenty First Century: Voluntary Action and the Global Agenda*, Kumarin Press, New York.

Leach, M. and Mearns, R. (1991) *Poverty and Environment in Developing Countries: An Overview Study*, final report to ESRC and ODA, IDS, Sussex.

Lees, C. (1995) 'Midwives say they murder female babies', *The Times*, 20.8.1995.

Leonard, H.J. (1989) *Environment and the Poor: Development Strategies for a Common Agenda*, Transaction Books, Oxford.

LeQuesne, C. and Clarke, C.A. (1997) 'Trade and sustainable development', in Dodds, F. (ed.) *The Way Forward: Beyond Agenda 21*, Earthscan, London, pp. 167–78.

Lindner, C. (1997) 'Agenda 21', in Dodds, F. (ed.) *The Way Forward: Beyond Agenda 21,* Earthscan, London, pp. 3–15.

Lipton, M. (1977) *Why Poor People Stay Poor: Urban Bias in World Development*, Temple Smith, London.

Mackintosh, M. (1992) 'Questioning the state', in Wuyts, M., Mackintosh, M. and Hewitt, T. (eds) *Development Policy and Public Action*, Oxford University Press, Oxford, pp. 61–89.

McCormick, J. (1995) *The Global Environment Movement*, second edition, Wiley, London.

McCully, P.O. (1991) 'Discord in the greenhouse: how the WRI is attempting to shift the blame for global warming', *The Ecologist*, 21,4, pp. 157–65.

Marray, M. (1991) 'Natural forgiveness', *Geographical Magazine*, LXIII, 12, pp. 18–22.

Mather, A.S. and Chapman, K. (1995) *Environmental Resources*, Longman, London.

Maxwell, S. (1998) 'Comparisons, convergence and connections: development studies in North and South', *IDS Bulletin*, 29,1, pp. 20–31.

Meadows, D.H., Meadows, D.L., Randers, J. and Behrens III, W.W. (1972) *The Limits to Growth*, Universe Books, New York.

Middleton, N., O'Keefe, P. and Moyo, S. (1993) *Tears of the Crocodile: From Rio to Reality in the Developing World*, Pluto Press, London.

Ministry of Housing, Physical Planning and Environment (1991) *Highlights of the Dutch National Environmental Policy Plan*, Department for Information and International Relations, The Hague.

Mitchell, B. (1997) *Resource and Environmental Management*, Longman, London.

Momsen, J.H. (1991) *Women and Development in the Third World*, Routledge, London.

Moser, C. (1995) 'Women's mobilisation in human settlements', in Kirkby, J., O'Keefe, P. and Timberlake, L. (eds) *The Earthscan Reader in Sustainable Development*, Earthscan, London, pp. 298–302.

Mosley, P. (1995) *Aid and Power: The World Bank and Policy Based Lending*, Routledge, London.

Munson, A. (1995) 'The UN Convention on biological diversity', in Kirkby, J. *et al.* (eds) *The Earthscan Reader in Sustainable Development*, Earthscan, London, pp. 55–62.

Myers, N. (1989) *Deforestation Rates in Tropical Forests and Their Climatic Implications*, Friends of the Earth, London.

Myers, N. and Myers, D. (1982) 'Increasing awareness of the supranational nature of emerging environmental issues', *Ambio*, 11,4, pp. 195–201.

*New Internationalist* (1990) *Global Warming: How to Turn Down the Heat*, no. 206.

—— (1997a) *Gene Dream*, no. 293, August.

—— (1997b) *Globalisation: Peeling Back the Layers*, no. 296.

O'Riordan, T. (1981) *Environmentalism*, second edition, Pion, London.

—— (ed.) (1995) *Environmental Science for Environmental Management*, Longman, London.

Patnaik, U. (1995) 'Economic and political consequences of the green revolution in India', in Kirkby, J., O'Keefe, P. and Timberlake, L. (eds) *The Earthscan Reader in Sustainable Development*, Earthscan, London, pp. 146–50.

Pearce, F. (1996) 'Squatters take control', *New Scientist*, 1.6.96, pp. 38–42.

—— (1997) 'The biggest dam in the world', in Owen, L. and Unwin, T. (eds) *Environmental Management: Readings and Case Studies*, Blackwell, London, pp. 349–54.

—— (1998a) 'Arsenic in the water', *The Guardian*, 19.2.98.

—— (1998b) 'On the poison trail', *The Guardian*, 26.2.98.

Pearson, R. (1992) 'Gender matters in development', in Allen, T. and Thomas, A. (eds) *Poverty and Development in the 1990s*, Oxford University Press, Oxford, pp. 291–313.

Peet, R. and Watts, M. (eds) (1996) *Liberation Ecologies*, Routledge, London.

Pepper, D. (1984) *The Roots of Modern Environmentalism*, Croom Helm, London.

—— (1996) *Modern Environmentalism: An Introduction*, Routledge, London.

Pickering, K.T. and Owen, L.A. (1994) *An Introduction to Global Environmental Issues*, Routledge, London.

Popham, P. (1997) 'Parents are paid to have the daughters India lost', *Independent*, 30.3.97.

Potter, R.B. and Lloyd-Evans, S. (1998) *The City in the Developing World*, Addison Wesley Longman, Harlow.

Potter, Robert B., Binns, J.A., Elliott, Jennifer A. and Smith, D. (1999) *Geographies of Development*, Addison Wesley Longman, Harlow.

Pretty, J.N. (1995) *Regenerating Agriculture: Policies and Practices for Sustainability and Self-reliance*, Earthscan, London.

Puckett, J. (1994) 'Disposing of the waste trade: closing the recycling loophole', *The Ecologist*, 24,2, pp. 53–8.

Reading, A.J., Thompson, R.D. and Millington, A.C. (1995) *Humid Tropical Environments*, Blackwell, Oxford.

Redclift, M. (1987) *Sustainable Development: Exploring the Contradictions*, Routledge, London.

—— (1992) 'Sustainable development and popular participation: a framework for analysis', in Ghai, D. and Vivian, J.M. (eds) *Grassroots Environmental Action: People's Participation in Sustainable Development*, Routledge, London, pp. 23–49.

—— (1995) 'The environment and structural adjustment: lessons for policy interventions in the 1990s', *Journal of Environmental Management*, 44, pp. 55–68.

—— (1996) *Wasted: Counting the Costs of Global Consumption*, Earthscan, London.

—— (1997) 'Sustainable development: needs, values, rights', in Owen, L. and Unwin, T. (eds) *Environmental Management: Readings and Case Studies*, Blackwell, London, pp. 438–50.

Reed, D. (ed.) (1996) *Structural Adjustment, the Environment and Sustainable Development*, Earthscan, London.

Rees, J. (1997) 'Equity and environmental policy', in Owen, L. and Unwin, T. (eds) *Environmental Management: Readings and Case Studies*, Blackwell, London, pp. 460–71.

Rees, W.E. (1992) 'Ecological footprints and appropriated carrying capacity: what urban economics leave out', *Environment and Urbanisation*, 4,2, pp. 121–30.

Rich, B. (1994) *Mortgaging the Earth: The World Bank, Environmental Impoverishment and the Crisis of Development*, Earthscan, London.

Rigg, J. (1997) *South-East Asia*, Routledge, London.

Rostow, W. (1960) *The Stages of Economic Growth: A Non-Communist Manifesto*, Cambridge University Press, London.

Sanchez, R. (1994) 'International trade in hazardous wastes: a global problem with uneven consequences for the Third World', *Journal of Environment and Development*, 3,1, pp. 137–52.

Satterthwaite, D., Hart, R., Levy, C., Mitlin, D., Ross, D., Smit, J. and Stephens, C. (1996) *The Environment for Children: Understanding and Acting on the Environmental Hazards that Threaten Children and Their Parents*, Earthscan, London.

Schwarz, W. (1991) 'They're not waving but drowning', *The Guardian*, 25.1.91.

Scoones, I. and Thompson, J. (eds) (1994) *Beyond Farmer First: Rural People's Knowledge, Agricultural Research and Extension Practice*, Intermediate Publications, London.

Sen, A.K. (1981) *Poverty and Famine: An Essay on Entitlement and Deprivation*, Clarendon, Oxford.

Seymour, J. (1996) 'Hungry for a new revolution', *New Scientist*, 30.3.96, pp. 32–7.

Shiva, V. (1989) *Staying Alive*, Zed Books, London.

Smith, P. (1995) 'Industrialisation and the environment', in Hewitt, T., Johnson, H. and Wield, D. (eds) *Industrialisation and Development*, pp. 277–302, Oxford University Press, Oxford.

Smith, P.M. (1991) 'Sustainable development and equity', in Smith, P.M. and Warr, K., *Global Environmental Issues*, Hodder and Stoughton, London, pp. 243–85.

Soussan, J. (1990) *Primary Resources and Energy in the Third World*, Routledge, London.

Starke, L. (1990) *Signs of Hope: Working Towards Our Common Future*, Oxford University Press, Oxford.

—— (ed.) (1997) *Vital Signs: 1996–97: The Trends that are Shaping Our Future*, Earthscan, London.

Stewart, (1995) *Adjustment and Poverty: Options and Choices*, Routledge, London.

Stock, R. (1995) *Africa South of the Sahara: A Geographical Interpretation*, The Guilford Press, London.

Tansey, G. and Worsley, T. (1995) *The Food System: A Guide*, Earthscan, London.

*The Ecologist* (1993) *Whose Common Future? Reclaiming the Commons*, Earthscan, London.

*The Guardian* (1992) 'The new waste colonialists', 14.2.92.

*The Guardian* (1998) '£500m blow in Blair's backyard', 4.9.98.

Third World Network (1989) 'Toxic waste dumping in the Third World', *Race and Class*, 30,3, pp. 47–57.

Todaro, M.P. (1997) *Economic Development in the Third World*, seventh edition, Longman, London.

Traisawasdichai, M. (1995) 'Chasing the little white ball', *New Internationalist*, no. 263, pp. 16–17.

Turner, R.K. (1988) *Sustainable Environmental Management*, Belhaven, London.

UNCHS (United Nations Centre for Human Settlements) (HABITAT) (1996) *An Urbanising World: Global Report on Human Settlements*, Oxford University Press, Oxford.

UNCTAD (1997) *World Investment Report*, United Nations, Geneva.

UNDP (United Nations Development Programme) (1985) *Challenge to the Environment: Annual Report*, UNDP, New York.

UNDP (United Nations Development Programme) (1993) *Human Development Report*, Oxford University Press, Oxford.

—— (1996) *Human Development Report*, Oxford Univeristy Press, Oxford.

United Nations (1989) *Prospects of World Urbanisation 1988*, UN Department of International Economic and Social Affairs, Washington.

UNRISD (1995) *States of Disarray: The Social Effects of Globalisation*, UNRISD, Geneva.

Vidal, J. (1998) 'Woman power halts work on disputed Indian dam', *The Guardian*, 13.1.98.

Visvanathan, N. (ed.) (1997) *The Women, Gender and Development Reader*, Earthscan, London.

Watkins, K. (1995) *The Oxfam Poverty Report*, Oxfam, Oxford.

Watts, M.J. (1983) *Silent Violence: Food, Famine and Peasantry in Northern Nigeria*, University of California Press, Berkeley.

WCED (World Commission on Environment and Development) (1987) *Our Common Future,* Oxford University Press, Oxford.

Werksman, J. (1995) 'Greening Bretton Woods', in Kirkby, J., O'Keefe, P. and Timberlake, L. (eds) *The Earthscan Reader in Sustainable Development*, Earthscan, London, pp. 274–87.

—— (ed.) (1996) *Greening International Institutions*, Earthscan, London.

Wilson, G. (1992) 'Diseases of poverty', in Allen, T. and Thomas, A. (eds) *Poverty and Development in the 1990s*, Oxford University Press, Oxford, pp. 34–54.

World Bank (1983) *World Development Report*, Oxford University Press, Oxford.

—— (1990a) *World Development Report*, Oxford University Press, Oxford.

—— (1990b) *Structural Adjustment and Sustainable Growth: The Urban Agenda*, World Bank, Washington.

—— (1992) *World Development Report*, Oxford University Press, Oxford.

—— (1993) *World Development Report*, Oxford University Press, Oxford.

—— (1994) *World Bank and the Environment, Fiscal 1993*, World Bank, Washington.

—— (1995) *Mainstreaming the Environment, Fiscal 1995*, World Bank, Washington.

—— (1997a) *Annual report, 1997*, World Bank, Washington.

—— (1997b) *World Development Report*, Oxford University Press, Oxford.

World Health Organisation (1990) *Public Health Impact of Pesticides Used in Agriculture,* WHO, Geneva.

—— (1992) *Our Planet, Our Health*, WHO Commission on Health and the Environment, Geneva.

WRI (World Resources Institute) (1990) *World Resources, 1990–91*, Oxford University Press, Oxford.

—— (1992) *World Resources, 1992–93*, Oxford University Press, Oxford.

—— (1994) *World Resources, 1994–95*, Oxford University Press, Oxford.

—— (1996) *World Resources, 1996–97: The Urban Environment*, Oxford University Press, Oxford.

Yearley, S. (1995) 'Dirty connections: transnational pollution', in Allen, J. and Hamnett, C. (eds) *A Shrinking World? Global Unevenness and Inequality*, The Shape of the World: Explorations in Human Geography series, no. 2, Oxford University Press, Oxford, pp. 143–82.

Young, E.M. (1996) *World Hunger*, Routledge, London.

# Index